FUEL FOR THOUGHT

An Environmental Strategy for the Energy Sector

Fuel for Thought

An Environmental Strategy for the Energy Sector

Prepared jointly by the Environment Department,

the Energy, Mining, and Telecommunications Department,

and the International Finance Corporation

This report has been prepared by the staff of the World Bank. The judgments expressed do not necessarily reflect the views of the Board of Executive Directors or of the governments they represent.

The material in this publication is copyrighted. The World Bank encourages dissemination of its work and will normally grant permission promptly.

Permission to photocopy items for internal or personal use, for the internal or personal use of specific clients, or for educational classroom use, is granted by the World Bank, provided that the appropriate fee is paid directly to the Copyright Clearance Center, Inc., 222 Rosewood Drive, Danvers, MA 01923, U.S.A., telephone 978-750-8400, fax 978-750-4470. Please contact the Copyright Clearance Center before photocopying items.

For permission to reprint individual articles or chapters, please fax your request with complete information to the Republication Department, Copyright Clearance Center, fax 978-750-4470.

All other queries on rights and licenses should be addressed to the World Bank at the address above or faxed to 202-522-2422.

LIBRARY OF CONGRESS CATALOGING-IN-PUBLICATION DATA

Fuel for thought : An environmental strategy for the energy sector /
   prepared jointly by the Environment Department, the Energy, Mining,
   and Telecommunications Department and the IFC.
      p. cm.
   ISBN 0-8213-4599-0
      1. Energy industries—Environmental aspects. 2. Energy consumption—
   Environmental aspects. 3. Renewable energy sources. 4. Sustainable development.
   I. World Bank. Environment Dept. II. World Bank. Energy, Mining, and
   Telecommunications Dept. III. International Finance Corporation.

HD9502.A2 F827 1999
333.79—dc21   99-052294

Contents

Acknowledgments

THIS PAPER WAS PREPARED JOINTLY by a team from the World Bank and the International Finance Corporation (IFC). The team was led initially by former Directors Andrew Steer of the Environment Department and Richard Stern of the Industry and Energy Department and subsequently by James Bond, Director of the Energy, Mining, and Telecommunications Department, Robert Watson and Kristalina Georgieva, successive Directors of the Environment Department, and Andreas Raczynski, Director of the Technical and Environment Department (IFC). Team members preparing the document included Joseph Gilling, Karl Jechoutek, Charles Feinstein, Odil Tunali Payton, Carter Brandon, Kseniya Lvovsky, John Strongman, and Richard Spencer (World Bank); and Louis Boorstin and Dana Younger (IFC). In addition, Harvey van Veldhuizen contributed the sections on the Multilateral Investment Guarantee Agency (MIGA). The Executive Summary was written by Mark Tomlinson (Africa Region) and Richard Ackermann (South Asia Region). Annex 2 draws extensively on a paper by Gordon Hughes. The Monitorable Progress Indicators matrix was produced in a collective effort by energy and environmental specialists from all Bank Regions, with overall coordination by Alastair McKechnie (South Asia Region). Editorial support on earlier drafts was provided by Daniel Litvin and Paul Wolman with support in desktop publishing by Carole-Sue Castronuovo and Vernetta Hitch.

Abbreviations and Acronyms

AIJ	Activities Implemented Jointly
APL	Adaptable Program Loan
ASTAE	Asia Alternative Energy Group
CAS	Country Assistance Strategy
CDM	Clean Development Mechanism
CEE	Central and Eastern Europe
CFCs	chlorofluorocarbons
CIS	Commonwealth of Independent States
CNG	Compressed natural gas
CODE	Committee on Development Effectiveness
CO_2	carbon dioxide
DALYs	disability-adjusted life years
DSM	demand-side management
DSS/IPC	Decision Support System for Integrated Pollution Control
EAs	Environmental Assessments
EASAG	East Asia and Pacific Energy and Mining Development Sector Unit
EC	European Commission
EER	Energy-Environment Review
EIPS	India: Power Environment Study
EMEs	Established Market Economies
EPA	Environmental Protection Agency (U.S.)
EPRI	Electric Power Research Institute
ESCO	energy service company

ESMAP	Joint UNDP/World Bank Energy Sector Management Assistance Program
ESSD	Environmentally and Socially Sustainable Development (network)
FI	financial intermediary
FINESSE	Financing Energy Services for Small-Scale Energy Users
FSU	Former Soviet Union
GCI	Global Carbon Initiative
GDP	gross domestic product
GEF	Global Environment Facility
GHG	greenhouse gas
GNP	gross national product
Gtoe	giga tonne oil equivalent
IBRD	International Bank for Reconstruction and Development
IDA	International Development Association
IFC	International Finance Corporation
IIASA	International Institute of Applied Systems Analysis
IPCC	Intergovernmental Panel on Climate Change
IPP	independent power producer
JI	Joint Implementation
kWh	kilowatt hour
LIL	Learning and Innovation Loan
LNG	liquefied natural gas
LPG	liquefied petroleum gas
MIGA	Multilateral Investment Guarantee Agency
MSW	municipal solid waste
MW	Megawatt
NEAP	National Environment Action Plan
NGOs	non-governmental organizations
NO_x	oxides of nitrogen
NO_2	nitrogen dioxide
OECD	Organization for Economic Cooperation and Development
OED	Operations Evaluation Department (World Bank)

OPC	Operations Policy Committee (World Bank)
OPs	Operational Policies
PCF	Prototype Carbon Fund
PELP	Poland Efficient Lighting Project
$PM_{2.5}$	particulate matter of 2.5 microns or less in size
PM_{10}	particulate matter of 10 microns or less in size
PSI	Private Sector and Infrastructure (network)
PV	photovoltaic
PVMTI	Photovoltaic Market Transformation Initiative
RAINS	Regional Air Pollution Information and Simulation
R&D	research and development
REEF	Renewable Energy and Energy Efficiency Fund
RPTES	Regional Program on the Traditional Energy Sector (World Bank)
SDC	Solar Development Corporation
SHS	solar home system
SIL	Sector Investment Loan
SO_2	sulfur dioxide
STAP	Scientific and Technical Advisory Panel (GEF)
TA	technical assistance
tbd	to be determined
TGs	Thematic Groups
toe	tonne oil equivalent
UN	United Nations
UNDP	United Nations Development Programme
UNEP	United Nations Environment Programme
UNFCCC	United Nations Framework Convention on Climate Change
USAID	U.S. Agency for International Development
USD	U.S. dollars
USDOE	U.S. Department of Energy
WBG	World Bank Group
WEC	World Energy Council

WHO   World Health Organization

WRI   World Resources Institute

The World Bank Group's fiscal year (FY) runs from July 1 to June 30 of the following year.

Preface

THIS PAPER FOCUSES on setting out an environmental strategy for the World Bank Group in the energy sector. It is aimed at two broad audiences. The first is internal: the World Bank Group's Executive Board, senior management, and sector staff. The second is external: the Bank Group's client countries, partners, nongovernmental organizations (NGOs), and the international community concerned with energy and environmental issues.

The paper has had a lengthy genesis to permit a broad participation of stakeholders. It was prepared by a team from the Bank's Energy, Mining, and Telecommunications Department and Environment Department. The team received input from the respective sector boards; from the Transport, Water, and Urban Department and the Operations Evaluation Department (OED); and from the Technical and Environment Department and Power Department within the International Finance Corporation.

An issues paper was written and discussed in the fall of 1996 with NGOs and representatives of industries and donor agencies. Consultations with Bank staff and the Subcommittee of the Committee on Development Effectiveness (CODE) were held during the preparation of a background discussion paper to provide a framework for the discussion and analytical details of the issues. The discussion paper was also posted on the World Wide Web, enabling an open-forum consultation with stakeholders over a six-week period in mid-1997. Comments from more than 50 organizations and individuals were received and posted for common access. In October 1997, OED completed a limited review of the Bank's implementation of energy and environment projects, providing critical additional input to the strategy paper. The paper was presented to the Executive Board in November 1998, and was subsequently posted on the Web for a second round of external consultations between December 1998 and February 1999. It was then further revised to incorporate the comments of the Executive Directors as well as those received via the Web from external stakeholders.

The paper is organized in four sections. Section 1 sets out the challenge arising from worldwide growth in energy demand and its potential impact on the environment. Section 2 reviews the World Bank Group's existing policies, strategy, and record. Section 3 outlines the new strategy, which contains both existing and fresh elements. Section 4 explains in detail how the new elements of the strategy will be implemented.

The paper mostly avoids jargon, but a few points on terminology may help. The energy sector, as defined here, encompasses the production, transportation, and conversion of primary sources of energy (e.g., hydropower, coal, gas, oil, and woodfuels) as well as refined petroleum products, electricity, and heat. The paper does not concern itself with all environmental issues, but is limited to those that can be addressed from within the energy sector. Similarly, it does not seek to provide comprehensive coverage of all energy issues, and is not intended to be a World Bank Group energy strategy. The environmental costs of pollution not reflected in the decision-making process are commonly described as externalities. A possible solution to pollution is to internalize these externalities through a variety of policy instruments, from environmental standards to energy pricing.

FUEL FOR THOUGHT

An Environmental Strategy for the Energy Sector

Executive Summary

Energy, Environment, and Development

ENERGY IS BASIC TO DEVELOPMENT. At the level of the individual, modern energy services can transform peoples' lives for the better. They can improve peoples' productivity. They have the potential to free millions of women and children from the daily grind of water and fuelwood collection, and through the provision of artificial lighting can extend the working day, providing also the invaluable ability to invest more time in education, health, and the community. They open a window to the world through radio, television, and the telephone. In the aggregate, they are a powerful engine of economic and social opportunity: no country has managed to develop much beyond a subsistence economy without ensuring at least minimum access to energy services for a broad section of its population. It is therefore not surprising to find that the billions who live in developing countries attach a high priority to energy services. On average, these people spend nearly 12 percent of their income on energy—more than five times the average for people living in OECD countries. As a "revealed preference," to use the economists' jargon, energy services are high on the agenda of the world's poorest people.

At the same time, the provision of energy services, especially those furnished through the combustion of fossil fuels and biomass, can have adverse environmental effects. In rich countries, much attention is directed to the regional and global consequences of fuel combustion, because many of the local effects of fuel combustion have been controlled—at considerable expense—over the past half-century. But in developing countries, the local environmental problems associated with energy use remain matters of concern that are at least as pressing as they were in industrialized countries 50 or 100 years ago. Furthermore, it is the poor who suffer most severely from such problems, because it is they who, through lack of access to better alternatives, are forced to rely upon the most inefficient and polluting sources of energy.

Environmental Concerns and Priorities for Developing Countries

There is no simple link between growth in energy consumption in a country and any associated trends in environmental damage. Greater use of fossil fuels means increased emissions of carbon dioxide (CO_2), which contributes to climate change and only indirectly affects human health. In most developing countries, however, it is high levels of exposure to the by-products of fuel combustion, particularly dust and soot, that is the more immediate problem. These local pollutants directly affect the health, life expectancy, and quality of life of anyone who is exposed to them at moderate or high levels.

In rural areas especially, the burning of traditional fuels in ill-designed stoves or hearths causes indoor air pollution that damages the health of women and children and exacerbates deforestation and land degradation. In cities, the burning of coal and other dirty fuels for household heating, for small-scale commercial business, and for industrial activity causes smog and acid rain. Urban transport—and especially poorly maintained buses, cars, and two-stroke vehicles—is a further, prominent source of local pollution. In developing countries, local air pollution causes as much as 4 million premature deaths each year. Nearly 60 percent of this total are the deaths of children under the age of five as a result of exposure to dirty cooking fuels. According to recent estimates, the economic losses caused by the billions of cases of respiratory illnesses—which result in reduced productivity and lower life expectancy, for example—amount to at least US$350 billion per year, or 6 percent of the GNP of developing countries.

Looking to the future, developing countries also have high stakes in global environmental issues. Coastal areas, including small island economies, that are prone to storm or flood damage are particularly at risk. Global climate change will potentially also harm human health (for example, by creating conditions for intensification of vector-borne diseases such as malaria and dengue fever), ecological systems (e.g., forests and coral reefs), water resources, and agriculture and fisheries. Developing countries, especially those of Africa, are the most vulnerable to these changes. But unlike local pollution, where the time constants are as short as months or even days, the time constants associated with climate change are very long (i.e., a century or more).

The economic costs of global climate change are expected to grow within the next 50–100 years, although the likely magnitude of the damages is highly uncertain. The World Bank Group will continue to support projects with global benefits wherever there are synergies with local environmental objectives and where the additional costs required to secure these global benefits are fully funded by international sources, such as the GEF. This is consistent with the Bank Group's commitment to support international conventions on global issues, such as the UNFCCC, and to assist borrower countries to meet their obligations under such conventions.

The World Bank Group's Strategy

This paper is intended to provide a strategy to guide staff and borrowers. It is not intended to change World Bank Group operational policies. The paper builds on the Bank Group's existing policies and activities and draws from lessons of experience. It involves three key instruments: policy assistance, knowledge management, and support for a variety of specific investments to anchor environmentally responsible policies and to support environmental best practice. Getting the policy fundamentals right is of the utmost importance. Knowledge exchange and investment—both public and private—need a sound policy framework within which to thrive; without the right policies in place, the sustainability of their environmental benefits is unlikely.

Promoting the Right Policies

This strategy emphasizes the early integration of country-specific energy/environment strategies into the planning of Bank Group assistance. It is our aim that priorities agreed with our clients—both bilaterally and in the context of a broader development framework—should lead to Country Assistance Strategies that will more fully address the policy issues associated with energy/environment concerns. This approach has three benefits: it clarifies the development constraints imposed by the negative impacts of energy use across sectors and social groups; it identifies the role that particular energy/environment initiatives can play in easing these constraints; and it leverages the effectiveness of corresponding initiatives in other sectors.

The specific policy areas in which we will seek to engage clients and stakeholders are:

■ Adopting a broad range of policies that target the principal sources of pollution across sectors, that are aimed at tangible improvements in environmental quality (e.g., of urban air quality) and/or performance (e.g., vehicle fuel and emission standards), and that balance the cost of compliance with environmental benefits. At the energy-environment interface, maximum benefits can only be captured by cross-sectoral policymaking. Given the fragmented nature of sectoral decision-making in most of the Bank's client countries, making progress in this area is a considerable challenge.

■ Developing policies for the energy sector that respond to the manner in which energy-environment issues impact the overall development objectives of the country concerned. Although the policy agenda will differ from country to country, energy sector reform, efficiency improvements, and provision of energy services in rural areas are likely to be important elements.

■ Accelerating the substitution of traditional fuels by modern energy and promoting new energy technologies, including renewables, by removing barriers to the development of their markets.

■ Strengthening monitoring and enforcement capabilities for mitigating the environmental impacts of energy production and use across all levels of government, with a focus on local government and an increased role of communities and civil society.

■ Promoting the restructuring of energy sector institutions and ownership as a focus of the energy-environment policy agenda, in order to capture the important environmental gains that energy sector reform involves. For example, pricing reform can enable proper reflection of the environmental costs of energy use and, along with market liberalization, can encourage improved energy efficiency. Likewise, competition in the provision of energy services improves reliability and gives an incentive to diversify and innovate supply, enabling clean technologies to compete on equal terms.

Managing Energy/Environment Knowledge

The generation and dissemination of relevant knowledge is an especially powerful tool in the effort to flex the link between demand growth for energy and more pollution. The World Bank Group aims to capitalize on its comparative advantage in two areas of energy/environment knowledge management. First, the Bank Group, in partnership with our clients and other stakeholders, will continue to support and extend research into the environmental impact of energy use by enabling more accurate assessment in different regions. Such work could include analysis of the impact of millions of small sources of pollution, such as cars and wood and coal stoves; the regional effects of acid deposition; and strategies to adapt to the effects of climate change. Second, the Bank Group will actively disseminate both new research and existing knowledge among our clients and partners. Access to the best knowledge of the issues will help our clients produce plans to efficiently target their particular environmental issues.

An important instrument in this respect will be the "Energy-Environment Reviews." Rather than simply extending project-based environmental assessments, Energy-Environment Reviews will take place well upstream of operations and will therefore help in the setting of operational priorities. The scope of the reviews will be adjusted to the needs and priorities of our clients, and will range from comprehensive regional assessments to the analysis of localized problems such as poor air quality in selected cities or the links between household fuel use, land degradation, and health in rural areas. Particular attention will be given to impacts on the poor.

Supporting Investments

In general, the World Bank Group will support investments in the energy sector of countries that have shown a commitment to improving sector efficiency through policy reform and/or sector restructuring. In certain cases, the World Bank Group may also make private sector investments in the energy sectors of countries that are in the early stages of reform, where such investments support the momentum toward increased efficiency gains through further sector reform. Priority will be given by the whole of the World Bank Group to supporting private sector investment made in parallel with the development of regulations that will level the playing field for investors and foster competition. At the same time, the World Bank Group's *Pollution Prevention and Abatement Handbook 1998* provides guidelines on good operating practices and maximum emission levels for power plants and other energy sector activities financed by the World Bank Group. These guidelines are being adopted by public and private sector financiers around the world. In addition, the World Bank Group Safeguard Policies will be applied and their compliance closely monitored to ensure that our operations do not have adverse social or environmental impacts.

In urban and peri-urban areas, Bank support for investments will generally focus on improving the efficiency of energy utilities and on promoting improvements in end-use efficiency. Wherever possible, this support will focus on the private sector, facilitating the switch from coal to gas or other cleaner fuels for households and small businesses and promoting better environmental performance of urban transport through improved fuel specifications and engine characteristics. In rural areas, investment support will focus on developing and mainstreaming new solutions to the challenge of expanding access to modern energy services, including renewables, and on more sustainable management and efficient utilization of traditional fuels. Special attention will be given to solutions that integrate new energy technologies in partnerships between local private investors and local communities. In many regions, these partnerships have the potential to supply energy services more effectively and in a more environmentally benign manner than is possible through utilities.

International trade in natural gas and electric power represents an opportunity to replace dirty sources with clean energy. The World Bank Group intends to explore and utilize all instruments, including guarantees, to support this trade.

Energy sector reform, supply- and demand-side efficiency improvements, innovative solutions to rural energy needs, and facilitating markets for cleaner energy technologies will be the key areas for the policy, knowledge transfer, and lending work that the Bank Group performs to assist our clients' efforts in combating both the local and the global impacts of energy development. Wherever possible, we will build on overlaps in these areas. The Bank Group will continue to play an important role in the transfer of knowledge and of financial resources, drawing on a wide range of partners, especially those in the private and nongovernmental sectors.

Implementing the Strategy

In each country, the World Bank Group's activities will start with the development of a clear understanding of the most effective strategies to achieve the fundamental objectives that have been agreed with our clients. Because client ownership of the strategies is essential to their successful implementation, many of the policy assistance and knowledge management initiatives will be local in focus. These will emphasize the provision of access to modern energy services, especially for the rural poor, and the mitigation of the impact of indoor and outdoor air pollution on health and productivity. Many investments will likewise have a national focus, but we aim also to strongly support those regional projects that would cut emissions by integrating supply systems.

Underlying this approach to implementation is our conviction that the best way to promote progress on global environment issues is to help our clients tackle issues of national priority—particularly those where the contribution to poverty alleviation and other development objectives is clear-cut and relatively immediate, and to which there is therefore strong local commitment. For example, programs of power sector reform are vital because they are key to both improving the financial and environmental performance of the energy sector and to freeing public sector resources for other urgent needs, such as education and health. At the same time, such programs help cut greenhouse gas emissions at their source through efficiency improvements and the substitution of cleaner fuels.

It is important to emphasize that local, regional, and global problems are not polarized into a set of "either/or" options. Many of the measures and policies required to reduce exposure to urban air pollution will also contribute to reducing emissions of regional pollutants, such as sulfur dioxide, and greenhouse gases, such as carbon dioxide. The extent of such overlaps varies from case to case. For example, a study of strategies to improve air quality in the city of Katowice, Poland, indicates that least-cost measures designed to reduce exposure to smoke and soot would achieve almost the same reductions in carbon dioxide as an alternative strategy focusing only on reducing carbon emissions—at no greater cost, but with much greater benefit for the local population. In contrast, a study of transport pollution in Mexico City and Santiago, Chile suggests that the global benefits of cost-effective options addressing this local problem would be limited. To return a significant global benefit, the study suggests that additional measures specifically addressing global considerations would also need to be taken.

There are also valuable opportunities for developing countries to combat regional and global environmental problems beyond "win-win" interventions alone. The World Bank Group aims to identify opportunities for further contributions to regional and global objectives that may be realized by expanding the scope of programs through the use of external resources—particularly those of the Global Environment Facility (GEF) or of innovative, pilot financing mechanisms such as the Prototype Carbon Fund (PCF) and private sector resources stimulated by the Kyoto Mechanisms—to promote sustainable development and reduce the marginal cost of greenhouse gas abatement in developing countries.

The flaring of natural gas provides an example of how the additional financial resources that have been made available to defray the costs of global environmental problems can be used. Natural gas, which is released as a by-product of oil production in a number of developing countries, is both a serious environmental problem and a tremendous economic loss—if burnt in a conventional gas turbine generator, the gas flared worldwide today could provide more than 3 percent of the electricity that is generated from all other sources. It does not appear financially attractive for oil producers to use this valuable by-product because of the imperfect market for the gas and because of the need to commit large amounts of money to pipeline or liquefaction infrastructure. The provision of financial assistance—justifiable by the global benefit—together with measures to improve market opportunities and contract security, could enhance to commercial viability programs to reduce flaring.

In the area of energy technologies, the World Bank Group–GEF Strategic Partnership for Renewable Energy, endorsed by the GEF Council in May 1999, proposes several models for expanding the scale and financial leverage of WBG-GEF renewable energy development efforts. These approaches seek to achieve a shift from individual project orientation to more effective, longer-term programmatic approaches. They include the use, where appropriate, of Adaptable Program Loans, long-term technology support mechanisms, an increased use of in-country financial intermediaries, and a more flexible decision authority for IFC investments.

Implementing the Strategy will require the training of current staff and the recruitment of new specialists. The Sector Boards for Environment and Energy, Mining, and Telecommunications will review the level and mix of staff skills needed to support the Bank Group's lending and non-lending services at the nexus of energy-environment and will make the necessary staffing decisions.

Monitoring Progress

Accurate monitoring of progress in implementing the strategy will be done using indicators that focus on the outcomes of interventions rather than on their inputs—i.e., on how much World Bank Group activities contribute to our clients' comprehensive development, rather than on how much we contribute to these activities. Outcome indicators are linked to programs of action and forecasts of short-term outputs, based on what the client is able to implement. This approach leads to strategic work programs designed to achieve environmental results that actually improve the lives of the poor.

A summary of the monitoring indicators is set out on the following pages. Detailed plans for each Bank Region are described in the annexes. Progress will be assessed regularly against these indicators to provide feedback to staff and managers and to enable fine-tuning of strategy. The World Bank Group will inform the Board of progress after one year, and will report formally in two years' time.

Monitorable Progress Indicators

FY 2000–02 Strategic Objectives	Outcomes	Actions Needed	FY 2000–02 Bank Outputs
Facilitate more efficient use of and substitution from traditional fuels in rural and peri-urban areas to reduce health damage from indoor air pollution and pressures on natural resources (land and forestry).	Significant progress in household access to cleaner commercial energy: ■ **Increase** share of cleaner commercial energy by 5–10 percent for at least five borrowers by the year 2005, and by 10–20 percent for at least 10 borrowers by the year 2010. ■ **Significant** increase in wood fuel production by increasing sustainable agriculture and land management. ■ **Substantial** increase in production and use of biogas and charcoal.	**Short and medium terms (FY 2000–05)** In at least eight borrowers, examine rural energy options and agree on the need for sustainable use of traditional fuels and reducing indoor air pollution; integrate these issues in CAS preparation; include traditional fuel use and rural energy access components in Bank projects; and facilitate government commitment to addressing the problems. **Long term (FY 2008)** Integrate energy access to rural communities and urban poor in Bank operations in about 10 borrowers and in other borrowers have firm government commitment to improved rural energy access; promote an approach focusing on the private sector, communities, and innovative financing for small local schemes.	Five traditional fuel activities/use project components. Bank operations in about 12 borrowers on rural energy (with renewable energy components for off-grid/isolated areas where appropriate). Review of Bank experience with traditional fuels and indoor air pollution. Consultation/dissemination workshops in at least four Regions. Sector reports for at least five borrowers that analyze the impact of traditional fuel use on health and the natural environment. Identify options (grid and decentralized) to improve access to commercial fuels in poor communities. Regional issues and options papers identifying interventions and specific targets to improve energy access for the poor. Implementation strategy, including local institutional strengthening, to address indoor air pollution problems and sustainable forest and land management.

FY 2000–02 Strategic Objectives	Outcomes	Actions Needed	FY 2000–02 Bank Outputs
Protect health of urban residents from air pollution due to fuel combustion in the residential, transport, industrial, and power sectors.	Measurable improvements in air quality in at least 20 major cities worldwide by the year 2010: ■ **Reduction** of atmospheric particulates concentration by at least 10 percent of current levels in 10 large cities by 2005, and by at least 20 percent in 20 large cities by 2010. ■ **Leaded gasoline** phased out in the major cities of half of borrower countries by 2005, and in all countries by 2010. ■ **Sulfur content** of motor diesel reduced to less than 0.5 percent in half of borrower countries by 2005, and in all countries by 2010.	**Short and medium terms (FY 2000–05)** Focus dialogue on reducing air pollution in a selected number of severely polluted cities. Facilitate dirty-to-clean fuel conversion of district heating and industrial boilers and individual household stoves and heating appliances by providing financing and work with governments toward the elimination of price distortions and commercialization/privatization of gas distribution and clean fuels and marketing. Assist governments in addressing the emissions problems arising from two-stroke engines. Facilitate the phasing out of lead and increasing the market share of unleaded gasoline worldwide through technical assistance. Facilitate the restructuring of the petroleum sector, improving fuel specifications and pricing policies. **Long term (FY 2008)** Have comprehensive air pollution control programs in place at least in 10 major cities and introduce similar programs in at least 10 other major cities.	Establish multisectoral teams to address urban air pollution involving energy, environment, transport, and urban staff. Sector work and other activities on lead phase-out and/or cleaner vehicle fuels/emission standards in at least four Regions. Preparation and/or implementation of about 15 dirty-to-clean fuel (e.g., coal-to-gas) conversion projects Preparation of clean air action plans for at least 10 cities (with city governments).

Monitorable Progress Indicators *(Continued)*

FY 2000–02 Strategic Objectives	Outcomes	Actions Needed	FY 2000–02 Bank Outputs
Promote environmentally sustainable development of energy resources.	Reduce air and water pollution and land degradation from the extraction, processing, and transport of commercial sources of energy by: ■ **Promoting** environmentally sound exploration and production of oil, gas, and coal. ■ **Reducing** spillage of oil, flaring of gas, and waste. ■ **Rehabilitation** and clean-up of selected degraded facilities and areas. ■ **Improving** utilization and disposal of by-products and residuals. ■ **Where applicable**, making progress toward compliance with the protocol on long-range transboundary sulfur dioxide and nitrogen oxide pollution. ■ **Increased** international trade in electricity, especially hydroelectricity, and natural gas.	**Short and medium terms (FY 2000–05)** Initiate dialogue and reach agreement with about 10 borrowers on the need for controlling the adverse environmental impacts of energy development; identify specific policies and investment operations; agree on a strategy with about 10 borrowers to meet identified outcomes and agree on specific policy measures and investment operations to meet these outcomes; and finance the closure of environmentally unsustainable coal mining operation in several large mines. Initiate and sustain dialogue on electricity and gas trade in at least three sub-regions. **Long term (FY 2008)** Implement strategy and launch several rehabilitation, clean-up, coal waste (ash) management, utilization, disposal, gas flaring and leakage reduction, and energy trade projects.	Agreed strategy and action plans for at least six borrowers. Preparation and/or implementation of about eight mining rehabilitation and restructuring projects. At least two projects or project components to address gas flaring. At least two projects or project components to improve the environmental performance of oil operations. At least one sector study/workshop on energy trade.

FY 2000–02 Strategic Objectives	Outcomes	Actions Needed	FY 2000–02 Bank Outputs
Mitigate the potential impact of energy use on global climate change.	For Bank-financed projects in at least 10 countries, achieve 5–10 percent reduction in cumulative GHG emissions projected for the year 2015 relative to cases without Bank Group financing. This expected outcome will be achieved through the implementation of a combination of: ■ **Power** sector reform and energy efficiency programs in 10 states or countries by FY 2008. ■ **Development** of cleaner sources of energy (e.g., hydropower, gas) and, where economically feasible, substitution of dirty fuels. ■ **Regional** integration of power grids. Increase the volume of energy trade between at least six countries by FY 2008. ■ **Doubling** of power generation through renewable energy sources in at least 10 borrowers by FY 2008.	**Short and medium terms (FY 2000–05)** Continue to build consensus on the need for power sector reform and promote the removal of energy price distortions; reach agreement with at least five governments (central or state) to implement pilot efficiency/DSM activities; foster ESCOs and strengthen relationships with strategic partners to develop alternative energy programs; agree with country/state governments to promote at least six renewable energy development projects; reach agreement in principle with at least six governments on the need for integrating regional grids and other links for promoting regional energy trade; and examine Bank support (lending, guarantees, advisory services) for clean energy projects and energy trade. **Long term (FY 2008)** Have in place policy, institutional, and regulatory framework to develop clean energy projects in at least 15 countries and develop an advanced dialogue with other governments; and mainstream advisory services to promote energy trade and renewable energy and energy efficiency projects.	About six power sector reform projects. At least 15 energy efficiency/DSM/renewable energy projects or project components (e.g., of power sector restructuring and rural electrification programs). Agreed strategy with at least four borrowers on promoting and implementing energy efficiency and renewable energy projects. Agreed hydropower and gas development strategies with at least four countries. Ten GEF-supported activities. Launch Prototype Carbon Fund (PCF) activities.

Monitorable Progress Indicators *(Continued)*

FY 2000–02 Strategic Objectives	Outcomes	Actions Needed	FY 2000–02 Bank Outputs
Develop capacity for environmental regulation, monitoring, and enforcement across all levels of governance.	Efficient environmental policies and regulations related to energy production and use; and transport fuels and vehicle emissions controls enacted and enforced by 2010 in at least half of client countries (significant progress in enforcement achieved by 2005).	**Short and medium terms (FY 2000–05)** In borrowing countries: Training/building local capacity for EAs; strengthening environmental regulations for private sector development (IPPs, etc.); developing monitoring capabilities; improving enforcement capacity of regulatory agencies and legal systems; supporting local-level governments and communities; and promoting private-public partnerships and participation. **Long term (FY 2008)** Establishing a coherent regulatory framework consistent with improved standards, regulations, and incentives for internalizing externalities, backed up by sound economic analysis of environmental impacts for planning and effective implementation programs.	Develop and implement an actions-oriented management training program for regulatory authorities. Environmental management capacity-building components in Bank energy projects, including power reform projects. Intensive advisory and consensus-building activities. Coordination and partnership with other donors; coherent outreach and communication strategy.

FY 2000–02 Strategic Objectives	Outcomes	Actions Needed	FY 2000–02 Bank Outputs
Make the Bank more responsive to addressing the adverse environmental impacts of the energy sector.	■ Develop more effective ways of generating and sharing knowledge; enhance awareness campaigns; build partnerships with strategic partners; and establish a framework for wider participation of NGOs and civil societies. ■ Full compliance with Bank safeguards policies. ■ Distill the policy content of existing good practice notes on the power sector and on energy efficiency into Operational Policies (OPs).	**Short and medium terms (FY 2000–05)** Improve the process of estimating ambient and emissions data related to Bank projects; share and disseminate information; launch awareness campaigns; identify champions of change in borrowing countries and promote wider participation of NGOs and civil societies; and facilitate partnerships with other donors and NGOs. **Long term (FY 2008)** Mainstream gathering and dissemination of knowledge; introduce environmental considerations into energy operations.	Review and reorganization of the thematic groups dealing with energy-environment nexus across PSI, ESSD, and HD Networks, and define the work program of these TGs according to the priorities of this strategy. Evaluation of staff skills and training and recruitment to fill skill gaps. Research and development on specific issues relevant to implementing the energy-environment strategy; e.g., economic analysis of environmental externalities. At least six Energy-Environment Reviews completed and made available. At least six more in progress. Periodic knowledge-management notes and workshops on energy-environment issues. Data on the environmental impacts of Bank energy operations are regularly assessed and made publicly available by the respective TG. Joint activities with UNDP, WHO, etc., and local and international NGOs on awareness- and consensus-building.

NOTES: 1) "Borrowers" refers to countries or sub-national entities (e.g., states, provinces, municipal governments) that are recipients of World Bank lending and non-lending services.

2) Estimates of numbers of borrowers implementing particular activities and numbers of projects and activities are provisional and are dependent on further analytical work, agreements with borrowers, and integration with Country Assistance Strategies and Business Plans.

3) "Dirty" fuels here means unprocessed solid fuels—coal, wood, and biomass—and heavy fuel oil; for example, fuels with high emission factors for particulates and other harmful pollutants. "Cleaner" sources of energy include renewables, natural gas, biogas, distilled petroleum products (e.g., LPG and kerosene), and processed solid fuels with lower emissions factors, such as smokeless coal or charcoal.

1. The Challenge

Energy and Development

Energy is vital to economic development. Without the fuels that power cars, trains, and planes, and without electricity, light, and heating, life in industrialized countries would be considerably less comfortable. In developing countries, however, it is not just a question of comfort. Poverty cannot be reduced without the greater use of modern forms of energy. Even now, around 2 billion people have no access to electricity, relying instead on traditional fuels such as dung and fuelwood. Those who are fortunate enough to have electricity spend an average of nearly 12 percent of their income on energy—more than five times the average for people living in OECD countries.

At the same time, the provision of energy services, especially through the combustion of fossil fuels and biomass, can have adverse effects on the environment. In rich countries, much attention is directed to the regional and global consequences of fuel combustion, because many of the local effects have been controlled—at considerable expense—over the past half-century. In developing countries, in contrast, local environmental problems associated with energy use remain at least as pressing as they were in industrialized countries 50 or 100 years ago. Furthermore, it is the poor who suffer most severely from such problems, because it is they who, for lack of access to better alternatives, are forced to rely upon the most inefficient and polluting sources of energy services.

As they seek to improve their standards of living, the challenge developing countries must confront is that of doing things differently from what has gone before. The economic growth of the industrialized countries and the higher-income developing countries was achieved in the past largely hand-in-hand with higher energy consumption. The increased supply and use of energy has led to more impacts both locally, in the form of urban air pollution and increasingly unlivable cities, contaminated water and ground, social disruption, and shrinking biodiversity; and globally, through massive increases in emissions of greenhouse gases (more details are given in Annex 1). While the industrialized world is now having to deal with the environmental legacy of its past energy use and policies, the developing world has the opportunity to flex the link between economic growth and energy consumption by pursuing efficient production processes and reducing waste; and at the same time to flex the link between energy consumption and pollution by relying more on renewables and by using fossil fuels more efficiently.[1] Only thus can the developing world achieve less energy-intensive growth and growth with less pollution.

The reality is that some tough trade-offs will need to be made, however. There will be cases where the benefits of increased growth and living standards will have to be weighed against increased pollution. This problem is further compounded by the social burden of increased pollution ("externality cost") not being shared equitably between the beneficiaries of economic growth—experience shows that the poor bear a disproportionate share of the costs of pollution. What is urgently needed is an analytical framework, beyond project-specific environmental assessments, within which the costs and benefits can be assessed and compared on a sector-wide basis, allowing the right choices to be taken early in the design of projects.

Rapid Growth in Energy Demand

In the last century, the industrialized countries consumed most of the world's commercial energy and accounted for most of the growth in energy demand. In this century, the developing countries are likely to beat them on both counts. The World Energy Council, for example, forecasts that on current trends world energy use will grow at 1.4 percent annually until 2020, but that growth in OECD countries will be just 0.7 percent

1. Through improvements in fuel quality and the introduction of cost-effective control technologies.

Figure 1.1 Primary Energy Consumption

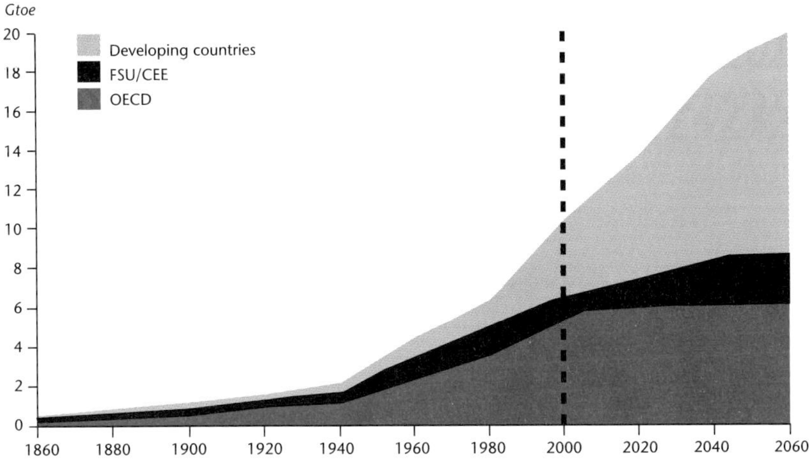

Gtoe

Developing countries
FSU/CEE
OECD

SOURCE: World Energy Council, World Bank.

The graph for the period 2000–2060 shows a scenario of future energy consumption based on current trends.

and growth in developing countries, 2.6 percent.[2] According to this scenario, the developing countries will match the total consumption of the OECD countries by 2015 (Figure 1.1). By 2050, they will have doubled it. But even then, the level of energy consumption per head of population in developing countries will be only one-quarter of that in OECD countries (Figure 1.2)—and the 2 billion rural people with no access to modern energy services currently use as little as one-twentieth of the energy enjoyed by even the poorest in the industrialized countries. Clearly, addressing the plight of the poorest demands that they consume more energy, not less.

Structural Changes in the Energy Sector

The broad trend toward increased energy consumption hides another, more complex series of movements that have been underway for the last decade and that are changing the backdrop of the entire sector. The way in which energy is delivered to the final consumer is undergoing a massive shake-up. First, energy production and delivery is moving from the public sector to private firms. Many governments are getting out of the

2. Global Energy Perspectives to 2050 and Beyond. "Mid-Range Current Trends Forecast of Energy Demand." World Energy Council. 1995.

Figure 1.2 Per Capita Energy Consumption in OECD and Developing Countries, 1990–2050, According to Ecologically Driven (Eco)* and Current Trends (Trends) Scenarios

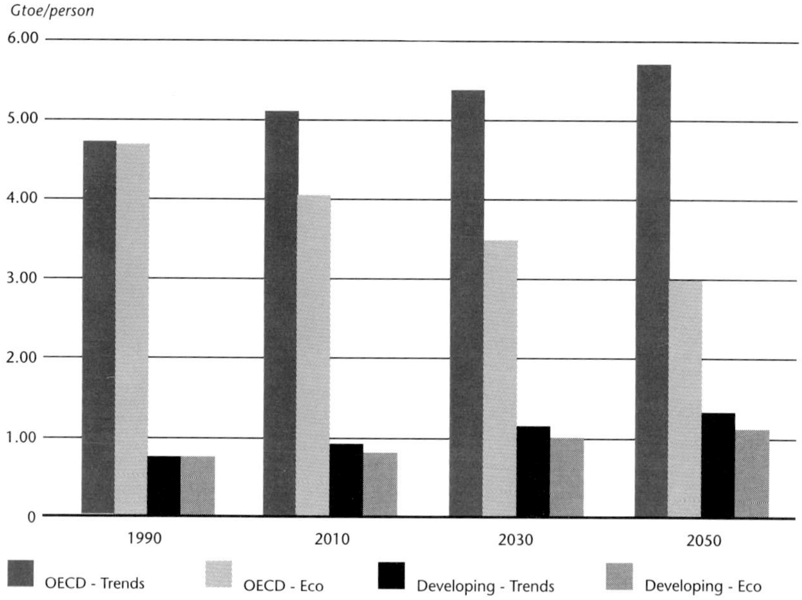

Gtoe/person

SOURCE: World Energy Council (WEC).

* The Eccologically Driven Scenario corresponds to best possible technology and full internalization of environmental externalities.

energy business, focusing instead on regulating the private operators who supply the consumers. Efficiency and environmental management are particularly sensitive to reform. Second, competition between private operators within sectors, and between different sources of energy, is becoming increasingly common. Vertically integrated monopoly providers are being replaced by a range of competing suppliers that are in continual search of better alliances and structures through which to deliver their energy services. Third, and perhaps most important, there has been a move away from central allocation based on planning tools to reliance on the market as a way of allocating resources such as investment finance, human resources, and fuel. This has profoundly modified the behavior of operators in the energy sector.

Moreover, together with the structural changes, there have been major technology developments. Combined-cycle gas turbines are today the technology of choice, displacing less efficient coal- and oil-fired steam turbines where gas is available; their lower cost offsets the higher cost of transport associated with natural gas. Turbine technologies have progressed to the point where smaller units are no more expensive per kilowatt than large turbines, leading to a modular approach whereby new tranches can be added over time. This is increasingly making distributed generation technically and economically more attractive. Renewables have also made inroads into conventional energy for off-grid energy services, and further declines in their cost will also contribute to grid-supplied power. The cost of photovoltaics, which for years were too expensive to be considered for anything other than remote, high-value applications such as emergency communications, has come down by an order of magnitude in a decade, bringing them within reach of more conventional uses, particularly in rural areas.[3]

Future Scenarios

During the 1970s and early 1980s, the period of the oil price shocks, there were widespread fears that major fuels were running short and that a world energy crisis was impending. But the world's energy industry—particularly the private sector—proved its ability to discover new reserves faster than the known reserves were used up. The future recurrence of political trouble in the Middle East could conceivably give rise to another oil price shock, but a major supply shortage looks unlikely for the moment. The far bigger worry now is the energy-environment nexus.

The most widely used fuels today are fossil fuels—coal, oil, and natural gas—all of which pollute the atmosphere to some degree. A range of views exists regarding the future share of fossil fuels in primary energy supply. According to World Energy Council projections, they will still account for almost two-thirds of primary energy even decades from now. Coal—perhaps the dirtiest of the fossil fuels—is too abundant and too cheap to be replaced easily. Despite the fears of two decades ago, oil is proving to be far more abundant than anyone had thought, and its convenience as a transport fuel is likely to guarantee it a place in future consumption. Natural gas is the cleanest of the fossil fuels, and substantial reserves have been found in many places, particularly in the developing world. The problem is that gas is not always located in the right place and

3. See *Rural Energy and Development—Improving Energy Supplies for Two Billion People.* World Bank. 1996.

Figure 1.3 Fuel Shares of World Energy Consumption

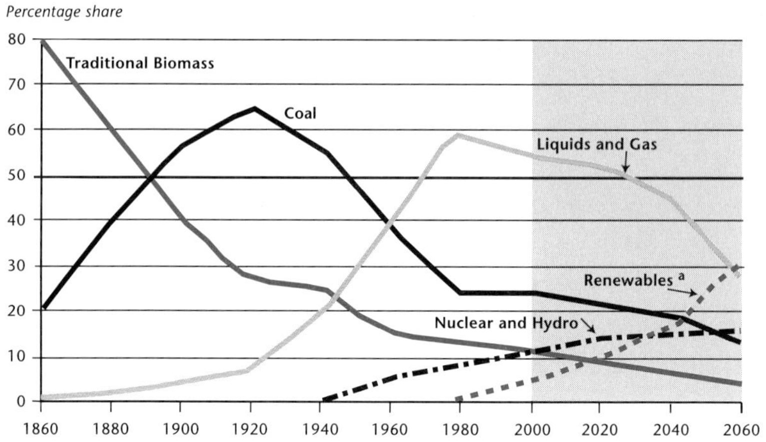

Percentage share

SOURCE: World Bank, based on WEC, Shell International, IPCC, and other scenarios.
NOTE: a. Renewables includes solar, wind, small hydropower, geothermal, and modern biomass.
The graph for the period 2000–2060 provides one possible scenario of future energy consumption that
shows renewables having the highest share. The scenario is not extrapolated from current trends.

that the investment required for its transport is very great, such that it is
viable only where it has a very large market. It should nonetheless be
noted that trade in natural gas has very positive environmental benefits
where it is used to replace coal or oil, and that trading of the resource is
increasing.

Some long-term scenarios, such as those envisioned by Shell International and the Intergovernmental Panel on Climate Change, postulate a
rapidly increasing share of renewable technologies. These comprise solar, wind, geothermal, modern biomass, and traditional hydropower generation. Under these scenarios, renewables could provide up to 50 percent of all energy by the mid-21st century, given appropriate enabling
policies and new technology developments. Hydropower—both traditional and small scale—is expected to play an important role in meeting
future energy needs. While it can offer the very positive global environmental benefit of reducing the use of fossil fuels, however, it can also
have negative local environmental and social impacts through requiring
the flooding of reservoir areas, thereby modifying the ecosystem, neces-
sitating population resettlement, and even enabling the introduction of
new diseases. The right trade-off must obviously be made. Figure 1.3

shows the possible fuel mix under a consensus scenario that reflects a pooling of different views.

Air Pollution and Its Dramatic Impacts

There is no simple link between energy consumption and air pollution. To date, most industrialized countries—which typically suffered from high levels of urban air pollution several decades ago—have managed to essentially clean the air in their cities, despite a very high and still growing energy use per capita. It is in developing countries, with far lower energy consumption, that local air pollution now is an immense problem. For example, exposure to smoke and soot has been recently estimated to cause as many as 4 million premature deaths each year, 40 million new cases of chronic bronchitis, and widespread cases of other respiratory illness. Significantly, nearly 60 percent of deaths are those of children under the age of five, the result of exposure to dirty cooking fuels because of the lack of access to modern energy. Deforestation is another problem afflicting the rural areas of many countries, due to the use of fuelwood.

Pollution by small particles is particularly harmful. Particles less than 10 microns in diameter, known as PM_{10}, are believed to cause cardiovascular disease, chronic bronchitis, and upper and lower respiratory tract infections. Another dangerous pollutant is lead from gasoline,[4] which has caused a permanent lowering of IQ in many children.

As well as harming human health, air pollution can damage ecological systems. Emissions of SO_2 and NO_x react with other chemicals in the atmosphere to form sulfuric and nitric acid, falling back to the earth mixed with rain. This acid rain—which sometimes falls thousands of kilometers from the original source of pollution—has been found to damage crops, forests, rivers, and lakes. Air pollution likewise has a detrimental effect on agricultural production, although there is much less clarity about just how and to what extent these effects occur. This is an area of current research.

While estimates vary, there is little doubt that the economic costs of air pollution are substantial (Figures 1.4 and 1.5). According to recent es-

---

4. Technically, tetramethyl and tetraethyl lead.

Figure 1.4 Air Pollution in Selected Cities: Total Suspended Particulates

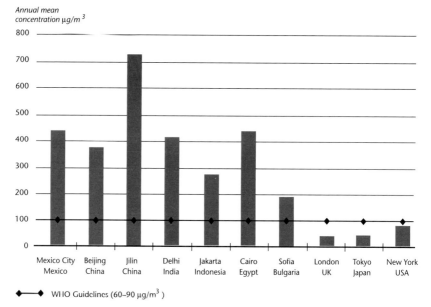

Annual mean concentration μg/m³

◆——◆ WHO Guidelines (60–90 μg/m³)

SOURCE: OECD Environmental data, 1995. WRI China tables, 1995. Central Pollution Control Board, Delhi. "Ambient Air Quality Status and Statistics, 1993 and 1994," *Urban Air Pollution in Megacities of the World.* WHO/UNEP, 1992. EPA, AIRS database. "Comparing Environmental Health Risks in Cairo." USAID, 1997. Europe's Environment: Statistical Compendium, 1995.

Figure 1.5 Health Costs from Exposure to Particulates in China

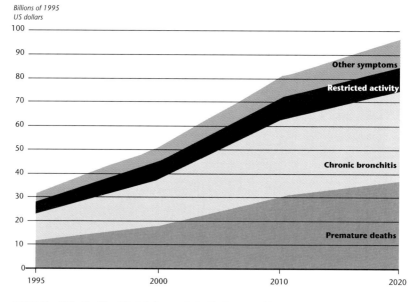

Billions of 1995 US dollars

SOURCE: *Clear Water, Blue Skies: China's Environment in the 21st Century.* World Bank, 1997.

timates, the economic losses due to the billions of cases of respiratory illnesses from indoor and outdoor exposure, which cause reduced productivity and lower life expectancy—amount to at least US $350 billion per
year, or 6 percent of the GNP of developing countries. In China, perhaps
the direst case, the cost of urban air pollution alone has been estimated at
5 percent of GDP in 1995. If today's trends continue, that figure could
reach around 13 percent of GDP by 2020.[5]

The Main Causes of Energy-Related Pollution

The burning of fuels is not always the main cause of air pollution. In
1997, forest fires in Indonesia, stoked by drought induced by the El Niño
phenomenon, caused a terrible smog across much of Southeast Asia.[6]
Public reaction to the fires in many developing countries demonstrated
the growing concern about air pollution. More often than not, fuel burning is nonetheless the main culprit. Pollution from fine particulates (PM_{10}
and $PM_{2.5}$), for example, generally comes from small-scale burning of
fossil fuels, cooking fires, cars, two-stroke engines, and other forms of
transport. It also comes from power stations and factories. Fuel burning
is also usually the main source of emissions of sulfur, nitrogen oxides,
and lead.

The precise cocktail of pollutants in the air varies from region to region. In many parts of the developing world, coal is still heavily used in
homes and in factories. Countries such as China, India, Poland, and
Turkey are now suffering the same pollution problems—urban smog
containing high levels of particulates and SO_2—that were suffered by industrialized countries before they cut back on coal use.

The burning of oil products—such as gasoline in cars and two-stroke
engines, and light oils used by households and small industry—results in
a different mix of pollutants. The mix may include, for example, lead and
volatile organic compounds as well as particulates and SO_2. This problem
is predominantly experienced by large cities in Brazil, Indonesia, Mexico,
and Thailand, but many developing countries also suffer jointly from this

5. World Bank, East Asia and Pacific Region. 1997. *Clear Water, Blue Skies: China's Environment in the 21st Century.*
6. The fires have been estimated by one specialist (Jack Rieley, University of Nottingham, U.K.) to have released more carbon dioxide into the atmosphere in six months than all the power stations and cars in Western Europe emit in a year. *Financial Times,* October 18, 1997.

type of pollution and severe pollution from coal burning. Two-stroke engines are particularly harmful, even in smaller cities like Kathmandu in Nepal. Whatever the case, any strategy to improve air quality clearly needs to address not just the larger sources of emissions, such as factories and power stations, but also the millions of small sources, such as cars and coal stoves.

There is some question as to whether the much higher profile of the private sector in energy has improved or worsened the situation as regards environmental damage. A competitive private sector operates energy facilities at demonstrably higher efficiencies than the public sector, resulting in lower primary energy requirements for a given energy service to the consumer and thus less wastage and a lower environmental impact. But at the same time, reducing emissions costs money, and competitive pressures sometime place operators—both private and public—in the position of having to make "cheaper versus cleaner" trade-offs. If the operator's responsibilities and requirements for compliance are not well defined, the result could be harm to the environment. These issues need to be reflected in environmental standards and regulations, which must in turn be monitored and enforced in order to create a level playing field in the energy market.

Potential Wastelands?

The environmental impact of modern energy in rural areas is often most significant at the production end of the energy chain. For instance, in their endless quest for new oil and gas reserves, many energy firms are moving gradually but inexorably toward environmentally delicate regions, such as offshore regions, arctic areas, and rain forests. Some energy firms will take advantage of lax enforcement of environmental standards in these areas; others will attempt to protect the environment using the best available technology. Either way, the firms will often find themselves operating in inhabited areas where their impact on the indigenous peoples can create difficult social issues that add to the already complex environmental concerns at hand.

Oil and gas pipelines in developing countries are often poorly maintained, and damaging spills and leaks are a constant hazard. Natural gas is primarily composed of methane, a much more potent greenhouse gas than carbon dioxide.[7] Small natural gas leaks are thus much worse, in terms of their potential impact on the global climate, than if coal or oil had been used instead of gas. Refineries, too, are often in serious need of upgrading to cope with demand for better-quality, cleaner products. Even where oil production and refining are carried out according to best practices, there is sometimes no market for the natural gas that is associated with many oil deposits. As a result, firms often flare off the gas, thereby emitting further carbon dioxide (Box 1.1).

Box 1.1 The Problem of Gas Flaring

Significant volumes of "associated" natural gas are routinely flared and vented in oilfields around the world. This practice is most intense in developing countries, although the vast majority of the population in these same countries has no access to commercial energy. There are also significant levels of gas flaring and venting in Russia and other former Soviet Union countries, but published data on the actual amounts of gas flared is hard to come by.

A variety of reasons account for the continuing practice of gas flaring and venting. Harnessing gas and transporting it to distant markets would require a considerably higher level of investment capital than would an equivalent oil development. Local capital markets in developing countries are often weak and unable to provide the needed capital for investments in gas utilization or export projects. At the same time, country risks, market distortions (including subsidies for competing fuels), and similar factors make it difficult to attract investment funds from foreign capital markets. As a result, the importance placed by cash-poor countries on generating revenue rapidly through oil exploitation tends to overshadow gas development needs and environmental concerns.

The 120 billion cubic meters of natural gas that is reported by Cedigaz to have been flared or vented in 1997 represents some 5 percent of total worldwide gas production. It is thermally equivalent (assuming a 35 percent conversion efficiency) to about 460 billion kWh of electricity, or approximately 3.5 percent of the annual electricity generated from all sources, worldwide. Natural gas is a non-renewable resource and a premium fuel. Its wastage on this scale is a major global issue.

Gas flaring and venting are of environmental concern because of their contribution to atmospheric emissions of greenhouse gases, especially carbon dioxide (CO_2) and methane, which contribute to climate change. Gas flaring and venting also cause some localized pollution and are strongly perceived in some localities to be responsible for a variety of medical conditions such as asthma, bronchitis, skin, and breathing problems.

---

7. Methane (CH_4) has more than 20 times the greenhouse effect of CO_2.

New coal mining developments can also cause environmental damage, such as the loss of forests and agricultural land and the potential pollution of surface water and groundwater. These consequences are typically manageable and remedies well-known, however; the most serious impacts of coal mining tend to be social, and concern its effect on the local population. Resettlement guidelines are now widely available, but much more work needs to be done to understand the effects of displacement of communities by large surface mines. Frequently, there is a lack of adequate consultation, compensation, and adequate provision of alternative housing or work. With hydroelectric dams, too, the loss of habitat and the displacement of people caused by the creation of the reservoir can create huge impacts and huge controversy. And, whether big or small, hydropower schemes can alter water quality and the patterns of water flow, which in turn affect aquatic species.

Global Climate Change

On top of these local and regional, urban and rural environmental problems comes global climate change. This is perhaps the least quantitatively understood, but also potentially the most devastating result of modern energy use. In 1995, the Intergovernmental Panel on Climate Change (IPCC), a panel of experts assembled by the United Nations, concluded after detailed scientific reviews that "the balance of evidence suggests a discernible human influence on global climate."[8] This human influence on the climate comes from emissions of three greenhouse gases (GHGs) in particular—carbon dioxide (CO_2), methane (CH_4), and nitrous oxide (N_2O). Such gases act like a blanket around the earth, trapping heat emitted from the earth's surface.

Even the more optimistic scenarios predict carbon emissions from burning fossil fuels (in the form of CO_2) to increase dramatically. They will probably double from today's level of 6.5 gigatons of carbon annually to 13.8 gigatons by 2050, according to mid-range IPCC scenarios.[9] Overall, about 80 percent of greenhouse gas emissions from human ac-

8. Intergovernmental Panel on Climate Change (IPCC). 1996. *Climate Change 1995: The Science of Climate Change.* Cambridge, U.K.: Cambridge University Press.

9. Intergovernmental Panel on Climate Change (IPCC). Leggett, Pepper, and Swart. 1992. "Emissions Scenarios for IPCC: An Update." In *Climate Change 1992: The Supplementary Report to the IPCC Scientific Assessment.*

Figure 1.6 Globally Averaged Surface Temperature: Historical vs. Projected Increase

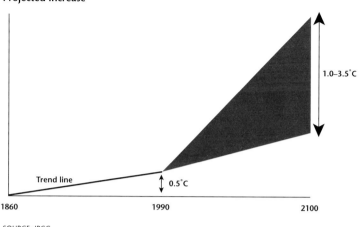

1.0–3.5°C

Trend line

0.5°C

1860   1990   2100

SOURCE: IPCC.

tivities are related to the production and use of energy—and particularly to the burning of fossil fuels. Most of the remaining 20 percent is associated with agriculture and changes in land use (such as deforestation).

Today, atmospheric concentrations of GHGs are higher than at any time in the past 160,000 years, and have risen substantially over the last 100 years. Global average temperatures have increased by 0.3–0.6 degrees Celsius over the past century, according to surface-based (land and ocean) temperature records. The IPCC predicts that, unless policies to reduce GHG emissions are widely implemented, global average temperatures will rise a further 1.0–3.5 degrees Celsius by the year 2100, with sea-levels consequently also rising (Figure 1.6). This would represent a rate of warming over the next century that would be greater than any that has occurred over the last 10,000 years, the period over which human civilization has developed. Global warming would adversely affect human health, ecosystems, agriculture, water resources, and human settlements, inter alia.[10] Compounding the problem, the impact of warming would most likely incur greater costs for developing countries than for industrialized countries: it has been estimated that a temperature rise of 2–3 degrees by 2100 would cost developing countries 5–9 percent of GDP. Is-

10. Intergovernmental Panel on Climate Change (IPCC). 1996. *Climate Change 1995: Impacts, Adaptation, and Mitigation of Climate Change.* Cambridge, U.K.: Cambridge University Press.

land states and low-lying countries such as Bangladesh would be particularly hard hit.

Industrialized countries are responsible for the bulk of the build-up of greenhouse gases in the atmosphere, accounting for about 80 percent of the present stock. Due to the long atmospheric lifetimes of most GHGs, equivalence in the stock contribution by developing countries to that of the industrialized world will not occur until approximately 2090. The developed world's commitment to cut emissions—made most recently at the Third Conference of the Parties to the United Nations Framework Convention on Climate Change in Kyoto in December 1997—reflects this historic responsibility.

Dealing with the climate change problem will eventually require a global effort to control GHG emissions. Current emissions from developing countries are growing fast, and by early this century are expected to exceed those of industrialized countries. As with other pollution problems, the focus of attention with regard to climate change will inevitably shift in coming years toward developing countries. The fundamental question, too, is the same: How do we reconcile economic growth, primarily fueled by coal, oil, and gas, with the need to protect the environment?

A Large Range of Remedies

It is worth noting that a range of remedies is available to deal with the challenge of pollution and poverty. Understanding the economic significance of each type of pollution is essential when considering intervention of any kind. Moreover, the relative importance to developing countries of each type of pollution is still uncertain. Annex 2 explores this issue in greater detail.

At one end of the range of remedies are reforms or investments that are economically sound in their own right, but which also provide environmental benefits. These win-win measures include strategies to improve the functioning of the energy sector as a whole through reform—and partial reform is generally better than none. Another is energy efficiency—getting more out for a given amount of energy in—which is attractive on both the supply and demand sides. Finally, there are cuts in subsidies for fossil fuels. Such subsidies are often short-term interventions to keep workers such as coal miners in jobs, but cutting them would not only save governments money and help the economy as a whole, but would also discourage excessive use of fossil fuels. The scope for these win-win measures is large: according to some estimates they could lead to 20–30 percent increases in energy efficiencies. But it is not always easy to take advantage of these opportunities: change can create losers as well as winners, and attention must be first paid to building the social and political constituencies that would support that change.

At the other end of the range are measures that, while good for the environment, are potentially costly. You win on the environment side, but you lose on another account. On balance, the benefits on the environment side—including better health and productivity—should more than offset the costs in terms of higher energy and product costs. These measures include investment in equipment to reduce emissions from factories and the closure of coal mines or fossil fuel power stations before the end of their natural life. Replacing vehicles powered by two-stroke engines would go a long way to reducing urban pollution in some cities, but would involve major investment to replace the rolling stock. Whether or not such measures are worth taking depends on how their cost compares with the benefit achieved in terms of reduced pollution. Unfortunately, such calculations require more data than is usually available in most developing countries. Moreover, while estimating the benefits of a reduction in local pollution problems is analytically difficult, it is still easier to accomplish than an accurate estimate of the more remote costs and benefits of reducing their contribution to the global problem of climate change.

Just as there is a large range of remedies and a vast variety of exacerbating circumstances, so too is the range of roles for the World Bank Group large and beset by many complicating factors. The WBG's role can range from awareness-raising and agenda-setting to a host of operations directly or indirectly targeted toward the environment. The complicating factors for the WBG are: (a) the extremely large range of issues, remedies, and manners of intervention in this area and (b) the comparatively small size and outreach of the WBG in the light of what are, after

all, some of the largest problems facing our planet for generations to
come.[11] Only by learning from the experiences of our clients and others
can we hope to have a real impact.

The following Section 2 reviews the World Bank Group's record so far.
Sections 3 and 4 set out the WBG's strategy for dealing with this challenge.

11. Historically, the World Bank has accounted for approximately 5 percent of foreign exchange
requirements and 3 percent of the total financing of our clients' energy sectors.

2. The Bank's Record So Far

STRIKING THE RIGHT BALANCE between energy development and protection of the environment is complex enough in theory, but even more challenging in practice. This section looks at how the World Bank Group has managed so far. Its conclusion is partly reassuring, partly worrying. The Bank is committed to wide-reaching reforms in the energy sector, which have been outlined in a number of recent papers. But although these have led to significant improvements in the financial performance of energy industries in our client countries, many of their environmental achievements have been disappointing.

The Bank's Activities in the Energy-Environment Area

Energy has been an important part of Bank Group work since the very beginning of the institution.[12] Today, Bank Group assistance in energy takes several forms. The first comprises regular Bank Group instruments such as loans, credits, guarantees, technical assistance, and advisory work (for IBRD and IDA); equity participations, syndications of commercial bank financing, and investment funds (for IFC); and MIGA political risk cover. Lending for energy typically represents between one-fifth and one-sixth of total annual commitments of the Group as a whole. The record is shown in Table 2.1.

But assistance in the energy sector goes beyond the Group's regular instruments. It also comprises a range of innovative programs including the Energy Sector Management Assistance Programme (ESMAP); the Global Environment Facility (GEF); the Asia Alternative Energy Group

---

12. The first energy loan, in an amount of $16 million to the Kingdom of Belgium to finance two steel mills and a power plant, was approved by the Board on February 8, 1949.

Table 2.1 World Bank Group Commitments in the Energy Sector *($ millions)*

	FY 1995			FY 1996			FY 1997			FY 1998			FY 1999 (Est.)[a]		
	Power	Oil, Gas, & Coal	Total	Power	Oil, Gas, & Coal	Total	Power	Oil, Gas, & Coal	Total	Power	Oil, Gas, & Coal	Total	Power	Oil, Gas, & Coal	Total
IBRD	1,803	462	2,265	2,459	613	3,072	1,613	414	2,027	1,115	1,472	2,587	385	303	688
IDA	434	142	576	348	46	394	276	22	298	874	15	889	119	23	142
IFC[b]	328	208	536	206	337	543	162	83	245	216	194	410	160	246	406
Total[c]	2,565	812	3,377	3,013	996	4,009	2,051	519	2,570	2,205	1,681	3,886	664	572	1,236

NOTES: a. FY 1999 estimates are as of June 25, 1999.
b. IFC data are for net approvals.
c. The total does not include World Bank Group guarantees. IBRD has completed six partial credit guarantees in China, Lebanon, the Philippines, and Thailand (total guaranteed amount is $673 million), and three partial risk guarantees in Morocco and Pakistan (total guaranteed amount is $491 million). IDA has issued one partial risk guarantee in Côte d'Ivoire for $30 million. In addition, MIGA has issued guarantees for energy projects in Argentina, Brazil, China, the Czech Republic, Honduras, Indonesia, Nepal, Jamaica, the Philippines, and Tunisia.
See also Annex 3.

(formerly ASTAE); the Clean Coal Initiative; the Africa-based Regional Program on the Traditional Energy Sector (RPTES); and the Solar Development Corporation (SDC), currently under implementation. Funding for these programs is shown in Table 2.2. These programs are intended to deliver tailored assistance to the Bank's clients supplementary to our traditional instruments, but in addition, they have been invaluable in helping us engage in agenda-setting beyond our own walls, to define our priorities in the sector, and to put together interventions that get to the heart of issues in the sector.

Table 2.2 Share of External Funding for Sustainable Energy Programs *($ millions)*

Program	External	World Bank
ESMAP (FY 1992–98)	6.3	2.3
GEF (FY 1991–98)	1.1[a]	3.4[b]
Asia Alternative Energy Program (FY 1992–98)	1.5	0.9
RPTES (FY 1992–98)	0.5	0.2
Rural and Renewable Energy Trust Fund[c] (FY 1998–99)	1.5	0.2
SDC (FY 1998–99)	0.4	0.4
Global Carbon Initiative (FY 1998–99)	1.2	1.7
Activities Implemented Jointly (FY 1997–99)	0.8	0.1
Total	14.3	9.2

NOTE: Where a range of years is indicated, the figure given is the annual average.
a. Administrative support from the GEF Secretariat for lending, supervision, and thematic support in GEF climate change projects implemented by the World Bank.
b. Including administrative support from the World Bank for lending and supervision in World Bank–financed components of GEF climate change projects.
c. Funded by Denmark for FY 1998–2000 for a total of $4.5 million external funding.

The oldest of these initiatives is ESMAP, a Bank-UNDP bilateral donors program created in 1983, shortly after the second oil shock. ESMAP was initially designed to produce quick, targeted technical assistance to help the Bank's client countries weather the disruptions caused by a quadrupling of the price of oil. Over the years, the quality of the program's analytical work has deepened as its focus has broadened, and it is no exaggeration to state that the make-up of the Bank's energy focus today is in large part due to the preparatory work done at the country level by ESMAP. Furthermore, ESMAP has for more than a decade been a conduit to bring energy sector staff into the Bank, such that today many of our energy staff are ESMAP alumni. ESMAP is mainly funded by bilateral donors (with a Bank contribution), and has been particularly successful in creating a cohesive donor community around its main areas of priority. These areas today are the linkages between energy and the environment; rural and household energy; renewable energy technologies; energy sector reform; energy efficiency; and international energy trade.

The World Bank is one of the three implementing agencies of the Global Environment Facility, which provides funding to developing countries and economies-in-transition for projects and activities offering global benefits. The GEF covers four important areas: reducing greenhouse gas emissions; supporting the phase-out of CFCs to eliminate ozone layer depletion; reducing marine pollution; and encouraging biodiversity. The GEF emphasizes removal of barriers to the implementation of climate-friendly, commercially viable energy-efficient technologies and energy conservation measures; the reduction of implementation costs for commercial and near-commercial renewable energy technologies; and the reduction of the cost of prospective low greenhouse gas-emitting technologies that are not yet commercially viable, with the aim of enhancing their commercial viability. The GEF has played an invaluable catalytic role by pioneering new approaches for Bank Group support of energy efficiency and renewables. It operates in part by helping developing countries bear the extra cost of measures designed to mitigate global environmental harm, by providing concessional funding and other incentives for environmentally favorable projects.

In 1992, the Bank and its donor partners established the Asia Alternative Energy Program to mainstream renewable energy and energy-efficiency options into Bank assistance to Asia. This cross-regional initiative (or ASTAE, as it became known) responded to the request of Asian countries and was intended as an operational implementation of an earlier initiative (Financing Energy Services for Small-Scale Energy Users; or FINESSE) undertaken by ESMAP. The program helps identify and implement commercially viable alternative energy projects and project components for Bank lending. The program has a strong client focus. It emphasizes Bank/GEF financing based on technical assistance, economic sector work, and dissemination of best practices to overcome technical, economic, and institutional barriers to environmentally sustainable energy services in developing countries. Located since 1997 in the East Asia and Pacific Energy and Mining Development Sector Unit (EASEG), the program continues its cross-support to the South Asia Region and is a notable model of successful Bank/donor partnerships. Since 1992, the Bank alternative energy portfolio in Asia has grown from a single renewable energy project to a seven-year portfolio of more than 30 projects in 12 Asian countries, representing more than $1 billion in Bank/GEF financial assistance (Annex 4).

The Bank recently launched a Clean Coal Initiative, an umbrella concept designed to encourage the use of more environmentally friendly coal technologies through sector work and in project design. This initiative is not intended necessarily to increase the amount of lending to the sector, but in several respects represents a new departure for the Bank: first, it examines the entire process, from mine face to ash disposal; second, it focuses on management as well as technology; and third, it concentrates on best industry practice using existing technology.

The Africa-based Regional Program on the Traditional Energy Sector (RPTES) is a bilaterally funded program to study the functioning of traditional energy markets in the Africa Region. The program has provided a unique glimpse into how these complex markets operate within the informal economy, and is expected to lead to new projects in Sub-Saharan Africa that will address informal energy markets and thus seek to improve access to energy for the very poorest.

The Solar Development Corporation (SDC) is an initiative launched by a partnership comprising the Bank, IFC, and a number of foundations. The objective of the SDC is to accelerate the use of solar photovoltaic (PV) systems through stand-alone, off-grid solutions delivered via sustainable private sector mechanisms. A consortium of fund managers has been selected and a management team is being assembled. The process of raising funds to capitalize the venture has also begun.

In addition to these programs, a number of other initiatives are currently under way:

■ The *Pollution Prevention and Abatement Handbook* was approved by the Bank's Operations Policy Committee and became binding on Bank Group operations in July 1999. The handbook provides guidelines for reducing air and water emissions from thermal power plants in developing countries and is expected to set a code of best practice for all industrial pollution.

■ The World Commission on Dams, chaired by the South African Minister of Water Affairs and Forestry, is designed to identify lessons learned and to develop a way forward to continue hydropower development as a viable source of renewable energy, with proper attention to both environmental and social issues.

■ The Activities Implemented Jointly (AIJ) Program, initiated in 1993 in collaboration with the Government of Norway, aims to investigate the potential for "joint implementation," by which governments or companies will contract with parties in another country to institute an activity that reduces greenhouse gas (GHG) emissions in that country. As a simple form of carbon trading, AIJ represents an important instrument for stimulating additional resource flows for the global environmental good.

■ The Global Carbon Initiative was also established to investigate the potential of market mechanisms in reducing global GHG emissions while supporting environmentally sustainable growth in developing countries. Under this initiative, the Bank is proposing to set up a Prototype Carbon Fund (PCF), whereby buyers and sellers of carbon offsets can participate in a pool of carbon-reducing investments. The fund, which was approved by the Bank's Executive Board in July 1999, will act as a market intermediary by obtaining capital from industrialized countries and private entities—which have binding emissions reduction obligations under the Kyoto Protocol (Box 3.3)—and investing it in projects that will result in carbon emission reductions in Bank-client countries.

■   The Energy-Environment Steering Committee is an advisory group of eminent persons convened by the Bank to provide independent advice on the transfer of environmentally sound technologies to the Bank's client countries.

The Bank Group's work in energy, both through its mainstream instruments and through these special initiatives, has been successful in achieving its objectives of providing cheap, reliable energy in developing countries. This paper investigates the degree to which these activities have also been successful on the environmental front. The paper first considers the principles at the heart of the existing strategy, and then examines regular Bank and IDA energy lending through an environmental lens to see how well we have done, using a review of a study prepared by the Operations Evaluation Department, a high-level audit group within the Bank. This reveals that there have been many failings as well as successes. The paper then looks at how some of the Bank's principles have been put into practice in the sensitive area of lending for coal. Finally, we review IFC's activities in the energy sector and its environmental record in this area.

Principles

This strategy builds upon a set of policies that have evolved over time and does not depart from them. In 1992, the Bank completed three important reviews of its experience with the energy sector in developing countries. The first review looked at the financial and economic performance of the electric power sector; the second addressed energy efficiency; and the third—whose conclusions were embodied in *World Development Report 1992*[13] —analyzed the relationship between economic development and the environment. The power sector and Energy-Efficiency Reviews culminated in Bank Policy Papers issued in 1993.[14]

The evidence from these reviews was stark: in a host of ways, govern-

---

13. World Bank. 1992. *World Development Report 1992: Development and the Environment.*

14. World Bank. 1993. *The World Bank's Role in the Electric Power Sector: Policies for Effective Institutional, Regulatory, and Financial Reform.* Also: World Bank 1993. *Energy Conservation in the Developing World.*

ments in developing countries intervene in energy markets with results that harm both economic growth and the environment. The reduction of these policy distortions would represent a win-win approach: the *World Development Report* pointed in particular to the need to eliminate subsidies for the use of fossil fuels and to make heavily polluting state-owned firms more competitive.

The *World Development Report* estimated, for example, that the cost of energy subsidies exceeded $230 billion a year in developing countries and transition economies. In fact, more than half the air pollution in the former Soviet Union and Eastern Europe appears to be due to these subsidies; their removal would not only sharply reduce local pollution but could cut worldwide carbon emissions by 10 percent. The report also noted that firms in polluting sectors, such as electric power generation, are often state-owned and state-run; and that the environment would benefit if these firms were made more accountable and exposed to competition.

The paper on power pointed to big inefficiencies in the electricity sector. Older power plants in developing countries, for example, consumed 18 to 44 percent more fuel for each unit of electricity produced than plants in OECD countries. Losses of electricity during transmission and distribution ranged between 20 and 40 percent. Although these losses included theft of electricity, they contrasted dramatically with figures of 8 percent in the United States and 7 percent in Japan.

The power policy paper set out a number of guiding principles for future Bank support of power sector programs. These principles are now also being applied in the Bank's operations in the oil, gas, and coal sectors.

Among them:

(1) In order to attract lending from the Bank for the power sector, client countries must make an explicit effort to reform and restructure the sector ("Commitment Lending").

(2) Regulatory processes must be established that are transparent and clearly independent and that avoid government interference in the day-to-day operations of power firms.

(3) In some of the least developed countries, the Bank will encourage the contracting out of service in the power sector to improve efficiency.

(4) The Bank will aggressively encourage private sector participation in the power sectors of developing countries—or at the very least, will encourage power firms to be run on a commercial basis.

The Bank's energy efficiency policy paper also set out some important principles. For example:

(1) The Bank will encourage client countries to improve energy efficiency by raising the issue as an important part of its country-policy dialogue.

(2) The Bank will be more selective in lending to energy supply firms, in particular avoiding those that are highly polluting or poorly performing. Governments should clearly demonstrate that they are putting in place structural incentives that will lead to greater energy efficiency.

(3) The Bank will identify and support measures to improve energy efficiency, both on the demand side and on the supply side. It will encourage the use of intermediation, in which specialist firms offer services to help improve the energy efficiency of other firms.

(4) The Bank will encourage the transfer of clean, energy-efficient technologies in its work in all sectors.

The policy papers were incorporated into the Bank's Operational Manual as GP 4.45 (Electric Power Sector) and GP 4.46 (Energy Efficiency) in May 1996. As part of operational policy reform, their content will also be included in OPs that will be issued in due course. Experience indicates that the guiding principles that underpin both GPs are robust. In applying these principles, the Bank has learned that countries vary widely in terms of their starting points, priorities, and resources, making the choice of tools and approaches more a matter of analysis and judgment than one of applying a uniform policy prescription. The Bank will continue to identify good practice in both power sector reform and energy efficiency, and will assess whether or not further guidance to staff needs to be defined.

Some Bank policies are applicable to all projects, including, for example, the newly revised safeguard series. Other recent Bank papers include a rural energy paper, *Rural Energy and Development*,[15] that addresses the

15. World Bank. 1996. *Rural Energy and Development: Improving Energy Supplies for Two Billion People.*

plight of the 2 billion poor people who do not have access to modern forms of energy. These people face a variety of environmental dangers, including indoor air pollution from cooking fires and the loss of tree cover and soils from unsustainable fuelwood harvesting. Many of these people are also vulnerable to resettlement associated with hydropower and coal mining projects.

Rural Energy and Development is an extension of the Bank's power and efficiency policies, and includes renewed commitments to:

(1) Extend modern energy supplies to unserved populations.

(2) Promote sustainable supply and use of biofuels.

(3) Introduce new and renewable energy technologies by:
- Promoting commercial pricing, in particular for oil products and coal
- Private involvement in distribution
- Providing incentives for extension of service
- Supporting agroforestry and biofuel programs
- Encouraging local initiatives and open markets.

The 1996 paper *Sustainable Transport: Priorities for Policy Reform*[16] also contains some recommendations relevant to the energy sector:

(1) Tackling health problems resulting from transport pollution is a priority. Standards and regulations, for example, should be used to eliminate lead and to severely reduce sulfur emissions; and policies on fuel pricing should encourage the use of cleaner fuels.

(2) The impact of transport projects on air pollution should be evaluated more carefully.

(3) Road pricing and/or taxation should be introduced whenever feasible and realistic, and should reflect all the domestic externalities of road use: not only air and noise pollution, but also road damage, congestion, and safety risks.

However impeccable they may sound, these principles will achieve little if they are not implemented in practice. Have they been?

---

16. World Bank. 1996. *Sustainable Transport: Priorities for Policy Reform*.

Experience in Policy Implementation

The most useful recent guide to the Bank's record in all areas of the energy sector is a study conducted in October 1997 by OED. Though not comprehensive, the study was substantial. It reviewed 98 projects that had been completed in the previous two years; 94 projects approved in FY 1994–97 that are now under implementation; and the 1998–2000 pipeline of projects.

Table 2.3 summarizes some of the findings of the OED study. The findings are reviewed in greater detail below.

Table 2.3 Energy and Environment in World Bank Projects *(FY 1994–97)*

	Number of Projects Addressing Strategic Issue	Share of Total (%)	Number of Non-Energy Projects Addressing Strategic Issue
Pricing and Restructuring			
Pricing	65	69	5
Restructuring	70	74	14
Supply-Side Efficiency	34	36	2
Billing and collection	19	20	1
End-Use Efficiency	20	21	4
Cutting Pollution			
Environmental policies and capacity building	22	23	11
Pollution abatement	18	19	18
Fuel switching	7	7	5
Renewables	10	11	4
Fuelwood and other	3	3	11

Pricing and Restructuring

Pricing and restructuring issues have been addressed in an impressive three-quarters of energy and non-energy projects. Some countries in Latin America, the Caribbean, and East Asia have made particular progress in pricing, and many other countries seem committed to making improvements. Credit is due not just to the Bank's tougher position on improving financial performance: the move to private investment has also been a factor. The decrease in oil prices has helped, and many countries have raised energy prices partly to attract private capital.

In the power sector, the Bank has given high priority to reform and restructuring, and generally projects will not be considered by Bank management without attention to these issues. Seminars with ministers and high-level government officials have often been pivotal in persuading governments to commit themselves to reform. This has been the case in China, India, and Sub-Saharan Africa.

We are making progress in implementing the reform agenda, but considerable work remains to be done. One problem frequently cited by Bank staff is the slow progress in the establishment of sound regulatory frameworks for the power sector. We are facing major problems still in South Asia, the Middle East, and Africa, where few countries have energy prices that reflect the long-run marginal costs of production; in most countries, prices remain to some extent distorted by cross-subsidization.

Supply-Side Efficiency

Supply-side efficiency is also an important issue. Power industries in developing countries often lose more than 20 percent of their electricity through theft or inefficiency. One way to stop this is to encourage private sector participation (as in Côte d'Ivoire) or complete privatization (as in Argentina and Chile). Losses in Argentina and Chile are now at an acceptable level of 10–12 percent. The progress in Argentina is also partly attributable to IFC loans to a now-private distribution company. Electricity metering can also help: in Tanzania, the installation of about 20,000 prepayment meters in households under a Bank-financed project not only improved collections but also resulted in a 5 percent drop in consumption.

OED nonetheless found that many of the recently completed Bank projects that aimed to cut electricity losses by publicly owned utilities achieved far less than hoped. Around 60 percent of projects achieved some loss reduction, but 80 percent did not meet their targets and half did not even come close, suggesting either that the targets were set too high or that there was a lack of commitment to achieving those targets. These poor results—together with the fact that electricity transmission and distribution firms are normally privatized later than electricity generators—suggest that progress in loss reduction will continue to be slow unless the Bank and its borrowers substantially increase their commitment in this area and accelerate the pace of private sector participation, including that at the distribution level.

There is huge scope for reducing energy losses in countries that use district heating systems, and the Bank has achieved some success in this area. Many district heating systems are hopelessly inefficient, particularly those in Eastern Europe and the former Soviet Union.[17] The Bank recently assisted a project to rehabilitate district heating in major Polish cities, and the improvements there are remarkable: 15–20 percent of energy has been saved, government subsidies—once 80 percent of cost—have been eliminated, and emission of pollutants has fallen by 15–20 percent. The Polish project was one of 12 conducted during the period FY 1992–97 that included district heating components. Over the next three years, we project our lending for district heating will increase substantially, with major projects lined up for Russia and Ukraine. IFC can also invest in district heating projects if a concession approach is used, as was pioneered in Poland.

End-Use Efficiency

The Bank's end-use energy efficiency programs in the electric power and other sectors, such as industry and district heating, are still at the pilot stage. Lending has increased over the last four years, but from a very low level. The Bank is currently engaged in 24 projects that include energy-efficiency measures, for which we are lending about $533 million. About 60 percent of this amount is earmarked for non-energy sector projects, including a $300 million loan for retrofitting apartment blocks in Russia. The project pipeline for FY 1998–2000 includes about 18 projects for an estimated $450 million in Bank lending. According to OED, the only ways for the Bank to realize significant energy savings from end-use efficiency measures in the coming years will be to intensify our efforts to develop pilot projects and to focus on a few large countries, such as China, Thailand, and Brazil. Projects are underway in Thailand and major operations are being finalized in Brazil and China that will include energy service companies (ESCOs).

An analysis of nine of the Bank's recently completed end-use efficiency projects shows that they have been mostly unsuccessful. Of these projects, six had no impact, because the end-use efficiency components were

---

17. Often these systems do not allow residents to control the amount of heat their houses or apartments receive. When the heating runs too high, people have no option but to open their windows.

not carried out by the government and other players, because energy au-
dits were not implemented, or because other measures taken proved to
be unsustainable. The impact of two projects is not known. Only one, in
Tunisia, had a positive impact on energy conservation. The projects
came up against a number of barriers: the lack of interest of consumers,
a lack of credit, and most notably, a low level of commitment from the
borrower. Borrowers tend to see demand-side management (DSM)
measures as risky from an institutional and policy perspective—they are
not familiar with the concept, and do not know how to organize, design,
and implement DSM programs. Borrowers have only been willing to undertake these components when grant financing has been available; and
in some cases, they have wanted to first test DSM on a smaller scale.

Much of the Bank lending for end-use in the 1990s was for industrial
energy conservation, and again these projects had mixed success. More
than 70 percent of the projects did not meet their objectives. Those that
did were mostly in the East European countries where access to foreign
exchange after the collapse of Communism allowed firms to buy new,
more efficient machinery.

Cutting Pollution

Only about 19 percent of energy sector projects approved in the last
four years included specific measures designed to reduce air pollution.
The figure for recently completed energy projects is 14 percent. In China,
Mongolia, Ukraine, and India, measures were aimed at curbing air pollution from coal. Other schemes focused on rehabilitating oil and gas
pipelines to stop leaks, such as the first and second Russia petroleum
loans and the China Sichuan gas loan; and on monitoring air pollution at
power stations. The Bank also financed the cleanup of oil pollution resulting from the rupture of the Komi pipeline in Russia, an oil spill worse
than that of the Exxon Valdez in Alaska. Outside the energy sector, six
projects are financing industrial pollution mitigation measures associated
with fuel combustion, and several other projects have included the installation of air quality monitoring systems. But though some investments
appear to showing success, there is little evidence to draw on as yet from
completed projects.

One exception is the World Bank's involvement in financing the Waigaoqiao thermal plant in China, which presents itself as a good example of lateral thinking about pollution control. The regulations of the local municipality of Waigaoqiao required that a special pollution control process—known as flue-gas desulfurization—be installed in the plant at a cost of $180 million. While preparing the project, the World Bank and the municipality agreed that because the plant was on the outskirts of the municipality, it would be better to spend instead $80 million on similar equipment for another plant closer to the urban center. This would not only save $100 million, but would also result in lower overall SO_2 emissions. Here, as in many other places, the best way to tackle pollution was to consider air quality across a whole airshed, rather than at a specific location.

Renewable Energy

World Bank lending for renewable energy is a new but growing activity. Lending for renewable energy over FY 1992–97 (excluding traditional geothermal and large-scale hydropower) amounted to $275 million in Bank/GEF loans, credits, and grants, leveraging total project costs in excess of $700 million. In Asia, the FY 1997 renewable energy project portfolio accounted for 8 percent of total power sector lending, up from 0.1 percent in FY 1992. Several factors contribute to increased Bank lending: technology improvements, cost reductions, utilization of locally available resources, fuel diversity, cost-effective applications, environmental awareness, and, most notably, increased demand by client countries.

The FY 1999–2001 pipeline of renewable energy components includes 20 projects (excluding geothermal), accounting for approximately $900 million of Bank plus GEF financing. These projects include freestanding renewables projects planned for Argentina, China, and India. The high proportion of lending in Asia reflects in part the efforts of the Asia Alternative Energy Program.

Activities in wind power to date have centered on India, where Bank participation has catalyzed more than 900 MW of capacity financed by the private sector. Government incentives have been key to spurring private participation in this environmentally friendly technology that reduces the need for fossil fuel-based capacity. The Bank is also working closely with China in the development of a grid-connected program that should in a few years deliver 200 MW of installed wind power.

In the PV field, the Bank is implementing projects in India, Indonesia, and Sri Lanka, and is developing projects in several other countries, including Argentina, Brazil, and China. These projects seek to overcome barriers to the deployment of solar energy systems and stimulate participation by the private sector.

In addition to PV and wind technologies, the Bank is working on projects in the areas of biomass, small hydropower, small geothermal, and solar thermal electric power generation. Bank support includes client-country assistance in project design and preparation, prefeasibility and feasibility studies, power purchase agreement/tariff development, technical specifications and standards development, equipment/system performance testing, and dissemination of best practices. Our future involvement in renewable energy projects will focus on completing existing projects, replicating these in current client countries and in new countries, expanding renewable energy projects and/or project components into non-power sector lending (e.g., agriculture, education, health, and forestry), and assisting countries to integrate renewables into their energy planning activities.

A Special Challenge: The Bank's Coal Lending

Environmental NGOs often ask why the Bank is lending to coal projects at all, given that coal burning is particularly harmful to the environment. But many developing countries would find it difficult to lift themselves out of poverty without the use of coal, which is a cheap and abundant fuel. Moreover, given the dramatic growth of private investment in developing countries, coal use is bound to increase dramatically with or without the Bank's involvement.

By staying involved in coal, the Bank's aim is to exert a positive influence on the whole coal chain, from the coal face through to the point of consumption and ash disposal, supporting projects that are both efficient and environmentally and socially sustainable. In a nutshell, this is the essence of the Bank's Clean Coal Initiative. Bank-supported coal projects are a small fraction of the total developing country investments in coal, but they can serve as a model of good practice. For example, Bank lending has had a large beneficial effect on the environment by shifting production from small, polluting installations to larger, more efficient plants. Such investments are often accompanied by stricter environmental standards and better compliance at the plant level.

In recent years, the Bank has notably lent for coal restructuring in Russia and the Ukraine (in FY 1996, total coal lending for these countries was $859.3 million). During the era of Communism, the coal sector in Russia and Eastern Europe was heavily subsidized, and this damaged both the economy and the environment. The Bank's lending has helped to reduce coal subsidies and also to cushion the impact of mine closures on jobs and on the environment. The Bank has also advised several Eastern European countries, in particular Poland, about reforming their coal industries and shutting down coal mines.

Similarly, the Bank has made loans to India, Mongolia, and China to reform their coal sector and to improve environmental performance. In India, for example, the Bank has not only supported the government's market-oriented reforms in the coal industry, but has also financed a project to help reduce the environmental and social consequences of new investments in mines, such as the impact on communities that must be resettled. In China, the Bank has helped the government by conducting studies on the efficiency and environmental impact of its coal industry and by financing, with the GEF, new clean coal technology (circulating fluidized bed boilers) for small industries and supercritical boilers for a new power plant.

IFC's Record

IFC's investments in the energy sector have traditionally focused on electric power generation fired by fossil fuels and on oil and gas production, coal mining, and motor vehicle manufacturing. In addition, many of IFC's investments in other sectors, such as chemical production, petroleum product refining, and cement manufacturing, have significant energy-use components. As with all IFC investments, these projects undergo extensive reviews to ensure compliance with environmental and social policies and guidelines, including those contained in the *Pollution Preven-*

tion and Abatement Handbook. IFC projects must also meet host-country environmental regulations.

Just like the Bank, IFC often acts in its developmental role as a catalyst in defining the scope of projects, and it can thus help to ensure that project designs are cost-effective in meeting environmental standards. However, private sector sponsors and lenders must always be satisfied that a project can provide a satisfactory risk/reward profile. As a minority investor that makes investments on a commercial (i.e., a nonconcessional) basis, IFC can induce private sector sponsors or lenders to make changes beyond those necessary to meet World Bank and local environmental guidelines only in those instances where the sponsors have identified a potential opportunity (win-win situations) or where IFC calls on third-party funding to defray incremental costs and/or risks (e.g., funds from the GEF or bilateral donors).

To date, IFC has invested on a purely commercial basis in several small and medium-sized hydroelectric projects (typically "run-of-the-river" projects), in one geothermal power project, one biomass power project, one solar equipment manufacturer, and in funds to support cogeneration. In addition, in FY 1997 IFC approved both equity and debt investments in the Renewable Energy and Energy Efficiency Fund (REEF). This is a pioneering investment fund of up to $200 million that addresses the need for an efficient vehicle to finance small and medium-sized sustainable energy projects. Although REEF will invest on a commercial basis, the sector is sufficiently new that IFC has secured up to $30 million in concessional funding from the GEF for direct investment in REEF's smaller and more innovative projects and to defray a portion of the fund manager's operating costs. Other examples of IFC's use of GEF concessional resources for sustainable energy projects include the Poland Efficient Lighting Project (PELP), which successfully promoted the use of compact fluorescent bulbs, and the Hungary Energy Efficiency Co-Financing Program, which uses partial guarantees to encourage commercial lending in this sector. In addition, the IFC/GEF Small and Medium-Sized Enterprises Program has made investments through intermediaries in energy efficiency, PV, and biomass projects, and the IFC/GEF Photovoltaic Market Transformation Initiative (PVMTI) is investing in PV projects in India, Kenya, and Morocco.

In the area of energy efficiency, IFC has invested in numerous projects that improve the efficiency of industrial energy use through the rehabilitation and upgrading of plants in energy-intensive sectors such as cement, chemicals, and pulp and paper. In addition to these process improvements, IFC has invested more directly in energy efficiency through several different types of projects, including (a) improvements in transmission and distribution equipment owned by private electric utilities; (b) the manufacturing of goods such as efficient light bulbs and insulation material; (c) profit-oriented energy service companies (ESCOs) that upgrade equipment and change processes to reduce energy consumption in client companies; and (d) financial intermediaries with credit facilities targeted at energy-efficiency improvements. To date, IFC has only made a few investments in each of these categories. However, an internal review that it commissioned in 1996 to assess the energy-efficiency sector identified a number of interesting opportunities and IFC is actively pursuing more investments in this area.

As indicated above, with the exception of energy-efficiency improvements associated with the rehabilitation and upgrading of industrial facilities, IFC has had limited experience with renewable energy and energy-efficiency investments. Furthermore, most of IFC's environmental projects in the energy sector have only been approved in the last few years, and many are not yet in commercial operation. As a result, a thematic evaluation of IFC operations would at this stage be premature. An exception to this situation is PELP, a GEF-funded project that has been completed. A final review of PELP's impact was completed by an external evaluator in June 1999. This evaluation concluded that PELP was effective and cost-efficient in promoting sustainable development through the increased use of compact fluorescent light bulbs. Similar evaluation studies will be undertaken for other IFC/GEF projects, and IFC will monitor and assess its own environmental investments as its portfolio of projects in this sector matures.

MIGA's Record

MIGA's (Multilateral Investment Guarantee Agency's) involvement in the energy sector consists of issuing contracts of guarantee for foreign direct investment in power generation and transmission systems, as well as in oil and gas production and transportation. As a provider of political risk insurance to foreign investors, MIGA does not participate in defining the scope of projects. However, like IFC, all applications for guarantee undergo extensive review to ensure compliance with environmental and social policies and guidelines, including those detailed in the *Pollution Prevention and Abatement Handbook,* and host country environmental requirements.

MIGA's portfolio includes contracts of guarantee to a small number of coal- and oil-fired power plants and to the expansion of an existing coal mine. MIGA has also been successful in facilitating several investments in renewable or cleaner-fuel energy systems (Table 2.4), including small run-of-the-river hydropower schemes, the privatization of existing hydropower systems, geothermal and gas-fired power generation, and gas distribution systems. Investors in these types of facilities often find MIGA's guarantee to be a critical factor in managing investment risk. In the area of energy efficiency, MIGA has provided guarantees that result-

Table 2.4 Outstanding MIGA Guarantees in Clean and Renewable Energy *(As of 18 June, 1999)*

Type/Country	Capacity	Number of Contracts
Gas-Fired		
Argentina	240 MW	2
Indonesia	500 MW	1
Pakistan	151 MW	1
Hydropower		
Bolivia	177 MW	1
Brazil	776 MW	2
Costa Rica	18 MW	2
Nepal	60 MW	4
Geothermal		
Guatemala	24 MW	3
Philippines	231 MW	1
Co-generation		
China	100 MW	2
Czech Republic	343 MW	2
Subtotal	**2,620 MW**	**21**
Other Power	**2,308 MW**	**27**

ed in co-generation in coal-fired and oil-fired power plants and improvements in electrical distribution systems, including interconnectors that improve system-wide reliability.

The Lessons for the World Bank Group

The OED review showed that private sector participation in the power, oil, and gas sectors has increased rapidly over the last few years (though for coal it is still early days). Putting the principles of reform into practice has had a clear and positive impact on the financial performance of the energy sector. The environmental benefits are less apparent. In many cases, governments have made insufficient efforts to either establish environmental regulations or to create the institutions that should enforce them.

For renewables and energy efficiency, the OED review highlighted that although the Bank's activities in alternative energy are relatively new, significant lessons learned in technical, policy, institutional, and financial areas are already being shared and applied in project preparation activities Bank-wide.

In the coal sector, the lesson is that the Bank can make a difference if it deals with the entire chain, from mine face to ash disposal, and if it uses best industrial practice. In the oil and gas sectors, where the private sector is more dominant, the Bank has been more selective but nonetheless continues to promote environmental and social safeguards. For the IFC, the lessons are that working with the private sector can induce efficiency improvements, but also that there are many specific opportunities for stand-alone energy efficiency projects.

Overall, the lessons from the Group's lending in the area of energy and the environment are threefold: that more time than initially estimated is needed to achieve results on environmental and social issues; that commitment is often missing on the part of the borrower to stay the course and to achieve real change; and that while there is strong engagement in the reform agenda, the strength of the Group's commitment to energy efficiency and the environment is not what it should or could be. The Group must substantially increase its efforts and improve its staff and skills mix if it is serious about implementing its principles in these areas.

3. ■ The Strategy

THE NEW STRATEGY aims not just to apply the lessons of the OED study, but to indicate clearly that, despite its relatively small size and outreach, the Bank Group aims to be in the forefront of the drive toward more sustainable use of energy.

This section provides an outline of the strategy to indicate how the different parts are articulated. Detailed recommendations are contained in the action plan in the following section.

Our Mission

The World Bank Group's client countries will consume increasing amounts of energy as they continue to develop. The Bank Group's main task is to help bring about a sustainable and rapid growth in incomes and to alleviate poverty. Within this process, our role is to ensure that energy is supplied at least economic cost and that it is used in the most efficient and sustainable way possible. Energy sector development has to take place in the context of a changing sector structure, characterized by rapid technological change and increased emphasis on the role of the private sector in sector operations and financing.

Capturing Win-Win Opportunities

The first step is to capture the win-win opportunities that, through economically attractive solutions, provide environmental benefits at no additional cost (Box 3.1). These opportunities include, at the very least, energy sector reform and restructuring; energy-efficiency improvements, both on the supply and on the demand side; and a switch to less polluting energy sources. The reform of the sector—by introducing competition, commercial principles, and private investment—provides the foundation for improving overall sector efficiency. Market prices give the right signals to producers and consumers alike, and alternative energy sources are no longer penalized. In a reformed sector, enterprises and households, whether urban or rural, will use energy resources more wisely. The World Bank Group will continue vigorously to pursue its policy of assisting reforms, and in general will engage only where such reforms are taking shape. In certain cases, the World Bank Group may finance efficient private sector investments in the energy sector of countries that are in the early stages of reform, and where such investments support the momentum toward further policy reform. Reform efforts in the energy sector will also be closely linked with the activities planned under the Private Involvement in Infrastructure Action Program, adopted in 1997.[18]

Economic and financial reforms alone, however, carry no assurance that their environmental effects will be exclusively positive. An environmental mandate and the norms that accompany the reform process must be clearly spelled out. If clear environmental standards are in place and effectively enforced, then and only then can institutional and regulatory reforms in the energy sector be considered unambiguously environmentally benign. Regulatory and legislative reform must include appropriate provisions for the environment.

Second, energy efficiency, the next major category of win-win opportunities, adds economically sound and financially attractive measures that are not always captured in the reform process. This includes rehabilitation of energy systems, operations and maintenance improvements, energy conservation investments, and improvements in energy-consuming processes that yield savings. The Bank Group will actively seek out these opportunities, and will encourage its public and private partners to explore profitable investments and enabling policies in this field.

Third, switching to less polluting energy sources can be both economically and environmentally sound. Switching to natural gas from conventional fuels can be a win-win solution if demand levels are large enough

18. *Facilitating Private Involvement in Infrastructure: An Action Program* was endorsed by World Bank Executive Directors in August 1997. It is designed to strengthen the WBG's ability to increase private participation in infrastructure within the context of its overall objectives to support poverty reduction and sustainable development.

Box 3.1 Win-Win Opportunities

On the Demand Side

- Improved customer billing and metering (electricity, gas, district heating) to link price and the rational use of energy
- Industrial boiler tune-ups
- Temperature and lighting controls
- Co-generation of electricity and heat
- Reduction of energy losses through building codes
- Optimizing water pumping by time-of-day tariffs, metering, and replacement of pumps
- Regulatory measures to remove obstacles

On the Supply Side

- Encouraging competition and private investment within a sound regulatory framework
- Cleaning up oil and gas leaks
- Improving coal mining and production
- Rehabilitating power plants and district heating systems, introducing loss-reduction programs in transmission and distribution
- Fuel switching to natural gas
- Large hydropower under the right conditions
- Gas trade (LNG; pipelines) and power trade
- Wind power, photovoltaics, biomass, geothermal, small hydropower

Obstacles to Win-Win Strategies

- Lack of access to financing
- Small absolute returns make efficiency measures less interesting for firms than big projects
- High levels of real or perceived risk
- Weak institutions and high transactions costs
- Inconsistent or ineffectual monitoring of energy savings over the lifetime of the investment

to cover the high infrastructure costs. New energy investments are generally more efficient throughout the supply chain and can replace outdated plant. Trade in electric power and natural gas provides additional opportunities to substitute clean energy for dirty sources. In rural areas off the electric power grid, renewable energy sources from solar or wind power are often the least-cost means to provide access to electricity, and are already replacing diesel, kerosene, or batteries for some 500,000 households. Elimination of lead from gasoline is often a good first step in the liquid fuels area. A multisectoral approach can follow, starting with monitoring and management and leading to a reduction of sulfur from mid- and heavy distillates, fuel switching (from motor diesel to compressed natural gas, for example), and overall fuel reformulation.

Operations that take advantage of the hybrid financial instruments available across the Bank Group and that leverage all the fund sources available (including concessional funding) are central to fostering win-win solutions. The Bank Group will increase its assistance for these solutions and its role as a global agenda-setter and champion of these alternative routes. It is a major aim of the Bank Group to make new renewable energy technologies and important underexploited sources, such as natural gas, increasingly win-win.

The Local Dimension

Win-win measures can go a long way toward reducing local environmental degradation, but are not in themselves a sufficient solution. The objective must be for all countries to integrate local environmental and social externality costs into either their energy pricing or investment decisions in order to minimize the total economic cost of energy production and use, including damage to human health and the environment (Box 3.2). The Bank Group must continue to play an important role in financing production of energy from fossil fuels, including coal, as well as production of electricity from hydropower. A too-rapid or over-selective shift to energy efficiency and renewables runs the risk of reducing the Bank's effectiveness upstream, its ability therefore to exert leverage in major issues, and ultimately its ability to ensure that the externality costs of these important sources of energy are adequately addressed.

The World Bank Group will only finance energy investments that respect environmental and social standards, that are economically and technically efficient, and for which there is a commitment to identify and meet externality costs in a practical and justifiable manner. Environmental standards are a proxy for the internalization of externality costs. In the absence of detailed cost/benefit information, good industry practice, as reflected in the *Pollution Prevention and Abatement Handbook,* is a practical

Box 3.2 Reducing Local Damages

Reducing local air pollution damages beyond the level of win-win implies that somebody—whether polluters, consumers, or society at large—must bear the cost. However, in most air-sheds with high pollution levels, the financial cost of abatement is small compared with the value of the health and environmental damage that can be avoided. For example, baghouse or electrostatic precipitators in thermal power plants and large industrial boilers are highly cost-effective, since particulates typically impose the largest health damages and the abatement cost per tonne is low. Policies to reduce pollution from small sources often promote cleaner fuels and improved technologies, and therefore overlap with the win-win approaches.

Overall, minimum air quality standards are increasingly common worldwide, and the cost of local compliance has not damaged the competitiveness of OECD countries or the many other countries that have adopted them. "Polluter pays" policies are also necessary to induce the development and market-ing of less polluting technologies. Without real incentives, technical change that favors cleaner fuels and technologies, renewables, and demand-side management will not happen.

There are three clusters of policies available to reduce local pollution damages: incentive-based poli-cies, regulatory (or command and control) policies, and public awareness policies. These are not mutu-ally exclusive, and combinations of policies often work best. Some represent more politically acceptable approaches, although at varying costs and efficiency.

	Price	*Quantity*	*Technology*
Incentive-Based			
Direct	■ Emissions charges ■ Product charges	■ Tradable emissions permits (within an airshed)	■ Technology taxes
Indirect	■ Fuel taxes ■ Performance bonds	■ Tradable input or production permits	■ Subsidies for R&D and fuel efficiency
Regulatory ("Command and Control")			
Direct		■ Emission standards	■ Mandated standards on control technologies
Indirect		■ Land use zoning ■ Bans and quotas on products and fuels	■ Efficiency standards for inputs or processes
Public Awareness		■ Timely public disclosure of ambient pollution levels; firm-level performance ratings; firm-level emissions; health and ecological impacts of pollution	

Emissions standards are typically the minimum requirement. Other incentives reduce the cost of com-pliance by encouraging flexibility, awareness, and innovative technologies. However, when moving be-yond win-win measures, careful analysis of relative costs and benefits is essential in helping countries pursue a least-cost path. While the application of environmental valuation techniques is in its early stages in Bank client countries, experience over the past 25 years in OECD countries has proven the rel-evance of these techniques to environmental decision-making. Assessment of costs and benefits can be based on a country's historical experience or on physical and economic models that trace the causative chain from emissions through to corrective expenditure or evaluation of lost output. While normally data intensive, rapid assessment methods for pollution management are now at an early stage of de-ployment in the Bank. Among such methods used in project preparation and Analytical and Advisory Activities work is the Decision Support System for Integrated Pollution Control (DSS/IPC). This tool per-mits rapid estimation of the extent and impact of pollution in a given situation, and provides support for decisions on pollution management.

means of cost internalization. In addition to this core business, the WBG will push forward on environmentally friendly projects and policies, while ensuring that they are also socially acceptable: clean coal programs, oil and gas leak and spillage projects, and urban transportation projects designed to reduce pollution, for example.

The Global Dimension

The World Bank Group recognizes the Intergovernmental Panel on Climate Change (IPCC) as the principal scientific authority on global climate change. It accepts the conclusions of the latest IPCC report that greenhouse gases emissions from human activities are affecting the global climate. It also believes that the consequences of climate change will disproportionately affect both poor people and poor countries. The WBG has an important role to play in helping to avert climate change, and it will assist its clients in meeting their obligations under the United Nations Framework Convention on Climate Change (UNFCCC). Under the 1997 Kyoto Protocol, some client countries with economies-in-transition have obligations to reduce their emissions of greenhouse gases; other clients—developing nations—have obligations to measure and monitor their GHG emissions, but not to reduce those emissions. In the case of developing country clients, the WBG will seek additional resources to ensure that they do not have to bear the additional costs of adopting climate-friendly technologies and policies unaided, and that their goals for national economic development and environmental quality are not compromised (Box 3.3).

Principles

In pursuing this mission, the World Bank Group will apply the following basic principles:

Principle 1: We will create a framework for environmentally sound energy sector development.

Creating an environmentally sound framework for energy requires continuing the push for sector reform that has been one of the main objectives of WBG operations in the past decade. The reform agenda has been driven by the need of borrowers to improve their financial perform-

Box 3.3 The UNFCCC and the Kyoto Protocol

The United Nations Framework Convention on Climate Change (UNFCCC) embodies the global response to the threat of climate change. The convention entered into force on March 21, 1994, and by July 1998 had been ratified by 175 countries. Many of these countries are Bank clients.

The ultimate objective of the UNFCCC is to achieve "stabilization of greenhouse gas [GHG] concentrations in the atmosphere at a level that would prevent dangerous anthropogenic interference with the climate system." Such a level should be achieved within a time frame sufficient to allow ecosystems to adapt naturally to climate change, to ensure that food production is not threatened, and to enable economic development to proceed in a sustainable manner. While adopting the "precautionary principle" as a basis for the implementation of potentially costly measures, the convention also recognizes the existence of an abundance of win-win policies and measures that benefit the global environment and the domestic economic and environmental interests of the Parties.

The convention recognizes that responses to climate change should be integrated with social and economic development with a view to avoiding adverse impacts on the achievement of national development aspirations. In addition, the convention is founded on the principle of "common but differentiated responsibilities," which confirms that all nations are responsible for the protection of the global atmosphere but which also recognizes that industrialized countries are accountable for the bulk of the present atmospheric stock of GHGs. Developing countries are least able to bear the costs of GHG mitigation, yet are most vulnerable to the effects of climate change. The principle of common but differentiated responsibilities is manifested most directly in the commitment of the Annex II (donor) Parties to provide new and additional resources through the convention's financial mechanism (entrusted to the GEF), and to transfer technologies on economically and socially beneficial terms. Additionally, the convention explicitly recognizes that developing-country energy consumption will need to grow for the achievement of sustained economic growth and the eradication of poverty, whereas Annex I Parties (OECD countries plus the economies-in-transition) should aim for the stabilization of GHG emissions. The convention does not impose mandatory emission restrictions on developing countries, but it does call on developed-country Parties to limit their anthropogenic emissions of GHGs.

The Kyoto Protocol. The Kyoto Protocol to the UNFCCC was adopted on December 11, 1997, and, if it enters into force, will result in binding carbon emission reduction limitations among 39 developed countries and countries with economies-in-transition (Annex B countries under the protocol). These parties agreed to ensure that their aggregate GHG emissions do not exceed their assigned amounts, "with a view to reducing their overall emissions of such gases by at least 5.2 percent below 1990 levels in the commitment period 2008 to 2012."

The Kyoto Protocol contains provisions allowing various elements of flexibility for Annex B countries in meeting their obligations, such as the ability to trade carbon reductions among countries ("emissions trading") and to jointly implement projects that could lead to carbon reduction on a project basis by reducing emissions or improving sinks ("joint implementation" or "JI"). Emissions trading can only take place among Annex B Parties, while Joint Implementation involving non-Annex B (i.e., developing country) Parties can take place under the Clean Development Mechanism (CDM), with crediting being allowed after the year 2000.

ance and to gain access to private sector funds. Sector reform and institution building ought also to lead to important environmental benefits, however. The environmental benefits of structural reforms come in many forms:

■ At the macro level, ensuring that energy prices properly reflect the long-run marginal cost of supplies has a powerful effect on total energy use in the medium and longer term, which will in turn be reflected in aggregate emissions of greenhouse gases. For example, the decline in energy consumption in Central Europe has been maintained despite the upturn in economic activity in Poland and elsewhere as users adjust to higher real prices for energy.

■ Competition, combined with appropriate price incentives, provides a strong basis for the adoption of new, cleaner technologies. The large-scale shift toward the use of natural gas for power generation, which has already brought large environmental benefits in the United Kingdom and is beginning to do so in countries as diverse as Brazil, Egypt, Hungary, and Thailand, has been closely linked to the deregulation of energy markets.

■ At the micro level, reliability of supply may be critical for many potential users. If electricity is only available for two to three hours a day and kerosene both heavily subsidized and rationed in supply, many farmers, firms, and households will invest in alternative sources of energy— sources that may be both dirtier and more expensive. Thus, diesel pumps and wood stoves are used in place of electric or kerosene appliances. Again, a primary benefit of structural reforms will be to give greater priority to quality and reliability of service, albeit at higher prices, enabling consumers to respond to real price signals.

To reinforce the potential environmental benefits of energy sector reform, some new steps are proposed:

■ **Sizing up the problem.** In order to assess the size of the problem in each of our client countries, we will create a new grant-funded technical assistance product, Energy-Environment Reviews, initially using resources from ESMAP and other trust fund programs.[19] Closely integrat-

19. Energy-Environment Reviews were initially outlined under the ESMAP program (see "Energy and the Environment: ESMAP after UNCED." *ESMAP Annual Report.* 1992).

ed with the CAS, the Energy Environment Reviews will cover local, regional, and global environmental issues at the country level relating to energy production, transport, and use. This sector work will identify priority actions and set a starting point for each country from which future improvements can be measured. For example, it will assess environmental impacts and opportunities for improvements along the coal chain, from production to end use, for countries that are large coal producers and users.

■ **Bringing environment and reform together.** Regulation is where reform and environmental management come together. To that end, the Bank's promotion of reform will focus on three pressing issues: (a) making clear and realistic choices about public vs. private responsibilities; (b) coordinating environmental standards with key reform objectives; and (c) improving the effectiveness of facilities licensing and standards enforcement by regulators.

■ **Setting environmental standards.** The setting of standards provides an important test of the consistency and effectiveness of regulatory frameworks. Many off-the-shelf tools can be used to cut the time and cost of institutional development in this area. The World Bank Group's *Pollution Prevention and Abatement Handbook* offers guidelines from which to start; more detailed, country-specific work will also be financed using a mix of externally funded ESMAP activities and Technical Assistance loans, such as the new Learning and Innovation Loans.

Principle 2: We will address local and regional environmental and social impacts as a first priority.

By far the greatest environmental cost of energy use today for developing countries is at the local level. The most common cost is health damage and shortened life expectancies, particularly due to respiratory diseases resulting from indoor and local air pollution, but there are other costs too: farmers may suffer from lower crop yields and decreased food production due to sulfur emissions that can be blown across an entire region; and social problems arise when populations are displaced by hydropower or coal mining projects.

Dealing with the local impacts of energy use has been a stated priority of the World Bank Group for a decade or more. We intend to become even more active in this area, in the following ways:

■ **Ensuring that environmentally sound investment decisions are taken.** We are beginning to systematically integrate local externality costs into the economic calculations relating to investments financed by the World Bank Group, although there is a dearth of country-specific methodological work. We will strengthen our work in this area in three ways: by refining the economic methodology we use, to ensure that options such as rehabilitation, energy efficiency, and airshed-based approaches are included; by providing training for Bank clients and staff through the World Bank Institute on how to carry out environmentally sound investment decisions; and by increasingly incorporating environmental values in the operations we finance. To reduce the adverse social and environmental impact of hydropower projects, we are participating in the World Commission on Dams to develop an international agreement on standards and measurements of best practice concerning resettlement and impact mitigation. Until the World Commission on Dams reports, we will continue to consider hydropower investment under its current social and environmental safeguard policies as they apply to hydropower development.

■ **Creating environmental best practices.** We will ensure that our energy investments use the most efficient technology available from an economic, environmental, and social point of view. In doing so we aim to establish best practices in terms of environmental requirements for other investments in the sector. We will accelerate our involvement in renewable energy, energy efficiency, and other environmentally sound energy projects. We will also explore appropriate instruments such as guarantees to support international energy trade, which often bring with them an environmental bonus.

■ **Dealing with the legacy. Cleaning up environmental damage.** Dealing with environmental problems resulting from past energy investments is a major issue in developing countries. We will accelerate the pace of rehabilitation and of the decommissioning of power plants, district heating systems, and refineries; of dismantling offshore oil production platforms; and of alleviating the effects of operations such as coal mining in instances where there is no clearly identified private owner. Environmental clean-up of this kind generates major benefits for the population in developing countries and economies-in-transition, but is seldom able to attract private financing because of concerns about environmental liability.

■ **Addressing small-source pollution.** As indicated in the review of Bank operations, pollution from small sources is often best addressed through non-energy sectors such as industry. In addition to further non-energy operations, we will design and implement targeted interventions in the energy sector to abate pollution from and to bring about behavioral changes related to small sources, including small industries and commercial establishments. Motor vehicles are a major and rapidly growing source of pollution in all urban areas, and we will be addressing them primarily through the transportation and urban sectors.

■ **Mitigating indoor air pollution: traditional fuels for cooking and heating.** This is largely a poverty- and gender-related issue, at the root of which is the lack of access to modern forms of energy for rural and poor urban households. Reducing exposure to indoor air pollution potentially has a large pay-off by improving environmental health for the rural and urban poor. Given the magnitude of the problem, a variety of measures are possible—for example, promoting the use of smokeless stoves and the greater use of LPG or kerosene in the near term. In the longer term, natural gas and renewable energy are more attractive options.

Principle 3: We will help tackle climate change.

The good news about global environmental problems such as climate change is that many measures imposed for other reasons—such as to reform and to restructure the energy sector or to reduce local pollution—also contribute to the solution of global problems. Many measures taken to reduce local pollution, such as the conversion from coal to gas, the use of renewable energies, and energy-efficiency improvements, simultaneously reduce greenhouse gas emissions. Designing the right response beyond such measures, however, is a very tough challenge: while the benefits accrue to the whole world, the costs are localized within the country where the investment or policy change takes place.

Therefore, as a first step, the World Bank Group will give priority to its policies, such as those on sector reform and energy efficiency, which achieve not just an improvement in financial efficiency but which also reduce greenhouse gas emissions. We will also continue in our key role as an implementing agency for the Global Environment Facility (Box 3.4).

Under the United Nations Convention on Climate Change, developing countries are not expected to finance by themselves investments that

---

Box 3.4 GEF Climate Change Operational Strategy

T he overall strategic thrust of GEF-financed climate change activities is to support sustainable measures that minimize climate change damage by reducing the risk, or the adverse effects, of climate change. The operational criteria for these GEF activities will be developed in accordance with this strategy and with GEF policies. The initial focus of GEF activities is on enabling activities that specifically support national communications. However, as the GEF builds on this foundation, the emphasis will switch to long-term measures, which will constitute the largest share of the GEF climate change portfolio. Enabling activities and short-term mitigation projects will constitute only a small share of the portfolio.

In line with the recommendations of the GEF's Scientific and Technical Advisory Panel (STAP), long-term programs will be developed to expand, facilitate, and aggregate the markets for the needed technologies and to improve their management and utilization, resulting in accelerated adoption and diffusion. The emphasis of these programs is three-pronged:

■ Removal of barriers to the implementation of climate-friendly, commercially viable energy-efficient technologies and energy conservation measures.

■ Removal of barriers to and reduction of implementation costs of commercial and near-commercial renewable energy technologies.

■ Reduction of the cost of prospective low greenhouse gas-emitting technologies that are not yet commercially viable, to enhance their commercial viability.

The following considerations will guide appraisals for short-term projects:

■ *Cost-effectiveness.* Cost-effective projects are those that mitigate a specified amount of greenhouse gas emissions for a given cost, typically a low unit abatement cost (approximately less than or equal to $10 per tonne of carbon).

■ *Likelihood of success.* When a project's funding is seen to be justified primarily in terms of the expected carbon abatement resulting from the project itself, it must have a high probability of success.

■ *Country-driven.* Proposed short-term projects must be country-driven and have the country's highest priority for funding.

deal with global externalities. Some investments may not be least-cost at the local level, even when all local externalities are factored in, but nonetheless may help reduce greenhouse gas emissions. In these cases, the WBG will seek concessional financing, such as from the Global Environment Facility, to cover the incremental costs to our client countries.

In addition to these measures, the Bank Group will increase its activities in two other areas:

■ **Using international market mechanisms to reduce the cost of carbon abatement.** The establishment of an international market for carbon emission offsets or credits should cut the cost of dealing with climate change, and was agreed in principle at the 1997 Kyoto conference on climate change. The WBG will help to develop this market.

■ **Long-term technological substitution.** In the long term, a shift to renewable and low-carbon energy sources will be needed to reduce CO_2 emissions. Given the long lead times required for the development and commercialization of new technologies, the importance of taking action now is evident. The WBG will start working with others—notably the GEF—to scale up the deployment of renewable and low-carbon energy technologies, in addition to increasing its funding for existing technologies such as natural gas development and transport.

IFC's Strategy

In view of the increasing commercial viability of renewable energy technologies, and in anticipation of potential changes in how the market values environmental externalities such as emission of greenhouse gases, IFC is actively investigating—and in certain cases, financing—environmentally friendly energy projects. In approaching newer technologies, IFC seeks out projects that fall into two basic categories:

■ Ventures that appear to be commercially viable from IFC's perspective but which are perceived as too risky by private sector investors and/or lenders.

■ Ventures that both IFC and the private sector consider to be close to, but not quite at, commercial viability.

To the extent that IFC's appraisal and structuring process can alter the market's perception of a viable project, IFC always prefers to develop projects where it can invest without additional assistance. In those cases where additional support is required and where there is a strategic or developmental reason to justify IFC's intervention, concessional funding sources such as the GEF or bilateral aid are sought. If such funds are accessed, IFC always tries to use financing mechanisms that are as close to commercial mechanisms as possible—for example, low-interest loans instead of grants—and to maximize the amount of commercial financing catalyzed by the concessional funds.

IFC's efforts to finance renewable energy and energy-efficiency projects go beyond being market-driven to reflect a commitment by IFC's senior management to expand investments in this area and to help accelerate market acceptance of environmentally sustainable energy projects. Moreover, IFC's Environmental Projects Unit was formed in 1996 to identify opportunities for new ventures by conceiving projects in-house, such as the Renewable Energy and Energy Efficiency Fund, and by seeking out potential sponsors of private sector projects. Potential sponsors are contacted through a concerted marketing campaign involving presentations by IFC at international conferences, and through direct contacts with leading companies, government agencies, and NGOs active in the sector. Once a potential project has been identified, IFC can provide specialized project development and financial structuring assistance. These extra efforts are necessary as IFC's experience has shown that most projects in this area are smaller, more time-intensive, and more difficult to execute than conventional energy projects.

In practical terms, IFC is actively pursuing investments in renewable energy and energy-efficiency projects, while also exploring opportunities in new energy technologies and activities related to climate change. IFC is also considering renewable energy projects, ranging from those that are often competitive with conventional energy sources, such as small-scale hydropower, modern biomass, and wind projects, to those that require some level of concessional assistance, such as photovoltaics. IFC is also assessing various types of energy-efficiency projects, including energy service companies, transmission and distribution improvements, and industrial upgrades. IFC's energy-efficiency strategy has benefited from a 1996 consulting study that identified a large potential market in this sector but which also cited many barriers, such as small project size or lack of collateral. New energy technologies being considered include fuel cells, "clean coal," and efficient vehicles. Finally, IFC is closely following developments in the area of climate change and is developing a strategy following the adoption of the Kyoto Protocol. This strategy includes identifying and implementing energy projects involving the purchase of carbon offset credits.

MIGA's Strategy

MIGA management is keen to support clean fuel and renewable energy projects. This interest is demonstrated in part by the relatively high percentage of such projects in MIGA's energy sector portfolio. MIGA's role as an insurer, however, places it in a unique role with respect to its sister agencies in that MIGA relies heavily on IBRD and IFC activities to foster a policy climate in the host country that is receptive to sound private sector investments in the energy sector. MIGA can then provide the political risk guarantees that encourage foreign investors to make the decision to invest.

Fulfilling the World Bank Group Mission

In translating the World Bank Group's mission into actionable items, it must be remembered that the WBG's funding is modest compared to the energy investment needs of developing countries. The Bank and the IFC need to expand their regular financing activities in the area of energy and the environment, but it will be necessary to go still further. The WBG in particular is well placed to draw on the experience of its clients to learn what is going on in the energy-environment area, what works, what fails, and how to overcome the obstacles. The Group can also use its convening power to share the results and lessons learned with the global community. Our action plan must include learning and sharing if we are to have an impact equal to our potential. Performance of this key role will be enabled by the Bank's new network structure and will be carried out by the Environment and Energy Sector Boards.

4. The Action Plan

THE PREVIOUS SECTION gave a broad outline of the World Bank Group's Environmental Strategy for the Energy Sector. This section explains in detail the measures that the Group will take to ensure that the new strategy achieves concrete results.

The WBG will make energy and environment part of its general business line in six main ways. It will (a) do more work further upstream in the project cycle and (b) identify how to bring new methods and technologies into the mainstream of lending operations. It will (c) improve standards of analysis of pollution problems, and will in particular monitor projects better, and it will communicate the results more effectively. It will (d) strengthen its capabilities to help address the new challenge of climate change and to (e) build up new strategic partnerships. Finally, to accomplish all of these actions, the Bank Group will (f) deepen its own skills.

Energy-Environment Framework: Working Further Upstream

To achieve the maximum possible leverage in improving sector efficiency and in building institutions, the WBG must work further upstream in the project cycle to identify those points where lending operations can produce the greatest results. As the Bank is a demand-driven organization, it is essential that the implementation of the strategy be firmly rooted in the Country Assistance Strategy (CAS) process, beginning with dialogue with governments and stakeholders on energy and environment issues and how these can best be addressed through WBG operations. It must also gather more information during the implementation of projects concerning their development activities. This can be achieved by building in project-level monitoring indicators and also through Analytical and Advisory Activities.

More Focus on Energy and Environment within the CAS Process; Inventories of Issues

Country Assistance Strategies are the principal documents setting out priorities for the World Bank Group within a country. It follows that energy and environment issues must be assessed first in that context. A brief review of recent CAS documents indicates, however, that discussion of energy sector issues tends to be sketchy. While being selective implies that not every CAS should address these issues, it is notable that only in larger countries, such as China, India, and Russia, are energy and the environment accorded high priority. With the exception of East Asia, little work also appears to have been done to assess the relative importance of urban air pollution compared with other health-related environmental problems such as poor sanitation and water quality.[20] A strategic approach is needed to address environmental issues in the energy sector for countries where these issues are significant. These strategies need to be based on rigorous multisectoral analysis such as that conducted for the Energy-Environment Reviews and similar studies.

Energy-Environment Reviews: A New Tool

Both the Bank's Operations Evaluation Department and its Environment Department have stressed the need to undertake Sectoral Environmental Assessments well upstream of lending operations.[21] To this end, we will undertake in the energy sector a client demand-driven program of Energy-Environment Reviews (EERs) and similar sectoral analysis.

---

20. *Can the Environment Wait? Priorities for East Asia* (World Bank, 1997) concludes that air pollution abatement investments in China have higher returns in health benefits. In other East and Southeast Asian countries, investments in water supply, sanitation, and abatement of air pollution have high but equal priority. Analytical work has been done in the Latin America and Caribbean and in the Europe and Central Asia regions to assess the impacts of urban air pollution, but comparisons with priorities in other sectors have generally not been made.

21. World Bank Operations Evaluations Department. 1996. *Effectiveness of Environmental Assessments and National Environmental Action Plans: A Process Study.* Also: World Bank. 1997. *The Impact of Environmental Assessments: A Review of World Bank Experience.*

This analytical work will help set priorities and lay the foundation for sustained follow-up. The activities will be upstream in nature, and not simply extensions of project-type environmental assessment. The range of activities will include:

■ *Full-Scale Energy and Environment Reviews.* These will be substantial pieces of country-specific sector work, covering a broad review of the environmental impacts of energy supply and use. They will use a systematic analytical framework and a participatory process to recommend a set of priority investments and policy reforms. Total budgets are likely to be in the region of $200,000 to $500,000.

■ *Rapid Energy/Environment Assessments.* These are envisaged as quicker country-specific, priority-setting reviews. They may consist of desk review of existing information, and some mission travel to identify and discuss with stakeholders the most pressing issues. Budgets may average some $100,000 each.

■ *Activities on Key Selected Topics.* This category supports narrower programs of action. Topics may include, for example:

■ Improved urban air quality management, or air quality management in other energy-intensive zones

■ Environmental clean-up strategies for energy supply sectors such as thermal power, oil production, coal mining and beneficiation, and oil refining

■ Environmental regulatory policy relating to energy supply and/ or use

■ Reduction of indoor and outdoor air pollution from household energy use

■ Quantification of environmental damage and health costs.

Establish and Apply Environmental and Social Standards

The Bank has a comparative advantage in being able to carry out a dialogue across many sectors and across many different ministries within a country. It must make use of this capability to encourage the development and implementation of the most cost-effective national air pollution standards. Governments will be encouraged to use the Bank's *Pollu-*

tion Prevention and Abatement Handbook to review the existing conditions against international standards and to prepare a set of pollution guidelines appropriate to local circumstances. In the interim, the handbook will provide guidelines for all projects undertaken in the energy and industrial sectors, including in particular all Bank and IFC projects. Assistance will be provided through the EERs and through follow-up technical assistance work to build institutions able to monitor and enforce the standards, to raise public awareness of the costs and benefits of environmental clean-up, and to help governments put in place pollution monitoring equipment. Through its involvement in the World Commission on Dams, the Bank will be proactive in defining environmental and social standards for hydropower development in dialogue with governments and NGOs.

Mainstreaming Environment in Bank Energy Operations

The Bank Group must make a greater effort to gain environmental leverage from its mainstream activities in the energy sector. Environmental priorities differ among countries and regions. Annexes 5 through 10 set out the strategic objectives for each region, the target environmental outcomes, and the actions and outputs needed to achieve them.

Eliminate Market Distortions

At the heart of mainstreaming environment within the Bank is the elimination of market distortions, particularly in energy pricing, by means of liberalizing production, distribution, and trade. As long as energy prices are subsidized or are not at the market level, and as long as gross inter-fuel pricing differences remain, it is difficult to formulate cost-effective measures to mitigate pollution from energy use. Unless such distortions are dealt with, the impacts of whatever else the Bank Group does will be limited. Tax policy, too, can be significant in both affecting people's behavior and as a means of internalizing externalities. In its macroeconomic and sector dialogue, the Bank will continue to press for the elimination of price distortions. Similarly, it will seek to assist clients in developing tax policies that are environmentally benign.

Finance Investments that Achieve
Environmental Outcomes

The World Bank Group will finance projects and programs that address indoor and urban air pollution, improve the environmental performance of the energy sector, strengthen environmental management, and produce global environmental benefits. Projects on power reform, energy efficiency, renewable energy, and cleaner fossil fuels are now being prepared in Brazil, China, and India. These can serve as models of best practice. The Bank's Photovoltaic Solar Home Systems project in Indonesia is also expected to provide a good example for other renewable energy investments. The Bank is starting clean coal initiatives in India and China and is supporting the Bolivia–Brazil gas pipeline, which is likely to increase significantly gas trade in Latin America. IFC is undertaking both the Renewable Energy and Energy Efficiency Fund (REEF) and the Photovoltaic Market Transformation Initiative (PVMTI). More such projects will be undertaken over the medium term in sector reform, energy efficiency, renewable energy, and the use of cleaner fossil fuels, such as natural gas. These projects will be closely monitored during implementation to allow the lessons learned to be incorporated in new investments.

Capture Environmental Gains through
Multisectoral Operations

A new look will be given to integrating energy considerations into non-energy operations, particularly through cross-sectoral, thematic teams. New ways of collaboration will lead to the design of integrated projects by including energy-related environmental considerations into urban development, urban transport, rural development, and poverty alleviation programs. Gains in energy efficiency are possible in water supply, sanitation, and irrigation and in urban transport projects. This includes encouraging the consideration of life-cycle energy costs in capital procurement decisions. The Bank's new Knowledge Management System (Box 4.1) is an instrument for raising awareness among staff of the potential for energy-related environmental activities.

Box 4.1 The Bank's New Sector Networks: Spreading the Knowledge

As part of the renewal process within the Bank, sector networks have been set up to improve the efficiency of operations, to deploy expertise more effectively, and to share state-of-the-art knowledge. Each sector network is governed by a board of managers and key sector staff who are drawn from all regional units. IFC management is represented on appropriate sector boards.

Sector networks have structured units headed by a Knowledge Manager who is responsible for the codification and internal dissemination of sector knowledge among staff working on the sector and to Country Directors.

Within the Energy, Mining, and Telecommunications sector, various Thematic Groups have been established. These groups, headed by senior sector staff who enjoy wide recognition as specialists in their field, help promote the emerging agenda of the Bank's operations.

Several of these groups address the environmental agenda, or specific parts of it. These include energy and environment linkage, energy efficiency, renewable energy, energy trade, sector reform, and cleaner fossil technology. The leaders of these groups are responsible for ensuring the efficient sharing of information and for pushing the environmental agenda into the heart of the Bank's operations. The Environment Sector Board is setting up similar Thematic Groups, some of which will link up with the appropriate groups from the Energy, Mining, and Telecommunications sector.

Use New Lending Instruments

A number of new lending instruments will be used to shorten the preparation time for innovative projects and to build up knowledge and experience. The new Learning and Innovation Loans (LILs) and Adaptable Program Loans (APLs) are well suited for operations, such as energy efficiency, renewable energy, and sector reform, that involve significant amounts of institution building.[22] One APL project has been approved in India for sector reform and a further two are planned; another one will be used for the proposed Energy Efficiency Project for Brazil. The IFC/GEF Renewable Energy and Energy Efficiency Fund provides simi-

---

22. Learning and Innovation Loans (LILs) are less than $5 million, can be approved by Bank management, and are designed to support small, time-sensitive programs to build capacity, to pilot promising initiatives, or to experiment and develop locally based models prior to larger-scale interventions. Adaptable Program Lending (APL) provides for funding of a long-term development program, starting with a first set of activities (the first APL loan) according to agreed milestones and benchmarks for realizing the program's objectives. The World Bank Board of Executive Directors approves the first loan and the long-term program agreement under which the full sequence of Adaptable Program Loans is prepared. Authority for approval of subsequent Adaptable Program Loans in the sequence will be with Bank management, subject to oversight and review by the Board.

lar opportunities to engage the private sector in these areas. The Bank's new lending instruments can be used to determine more specifically the priorities for energy-environment lending as a follow-up to technical assistance and Analytical and Advisory Activities.

More Focus on Rural Energy and Indoor Air Pollution

A specific action plan has been set out in the 1996 paper, *Rural Energy and Development*. The plan calls for local ownership and commitment and also for greater efforts on rural energy development within the World Bank. These proposals will be addressed through policy discussions on the issue with client countries, and through broadening the knowledge base within the Bank. A number of initiatives are already poised to increase the availability of renewable energy technologies in rural areas. These technologies have significant potential to improve access to clean energy. Examples are the IFC/GEF REEF, the IFC/GEF PVMTI, and the proposed Solar Development Corporation (SDC; Annex 4).

Expand Investment in Private Sector Projects

IFC is actively considering new investments involving small hydropower, geothermal, biomass, wind, and solar technologies, using both its own resources and funding from concessional sources such as the GEF. New GEF projects under implementation include an Efficient Lighting Initiative, which will disseminate to other countries the lessons learned from the Polish project. In the longer run, IFC is also investigating potential GEF projects that involve large-scale renewable energy facilities, advanced technologies such as fuel cells and "clean coal" plants, and efficient motor vehicles. IFC is additionally taking the lead for the Bank Group in implementing the SDC.

Improving Analysis, Communication, and Monitoring

Better Analysis of Local and Regional Externalities

The internalization of externality costs is a straightforward concept, but its implementation is less so. The Bank's ability to apply the principle of internalizing externalities depends upon how much is known of the damage functions of various local and regional pollutants. For example, estimates of health costs due to particulates have been made in a number of cities, but these are usually based on incomplete data or on the extrapolation of results from other countries.[23] The environmental costs of acid rain are known with even less certainty. As a result, governments are reluctant to spend money for environmental measures, especially in instances where this would reduce the pace of expansion of energy supplies. To ensure estimates of externality costs are incorporated as standard practice within project appraisal, further work is needed to supplement the existing practice of using estimates or ranges. This work should aim to quantify the cost of environmental externalities associated with different uses of energy and to enable an understanding of how costs vary according to geographical location. Work begun under the RAINS-Asia project, for example, continues to assess the extent of the projected high levels of sulfur deposition in Asia and to deepen understanding of the potential impact of acid rain on agriculture, forests, and lakes.[24] Research upstreaming of operations will increasingly include analysis of social and environmental externalities. Upstream work will also help identify pilot activities and requirements for regulatory capacity building.

Guidelines in the Bank's *Pollution Prevention and Abatement Handbook* provide a way to begin internalization of the externality costs of pollution. The emissions standards suggested for power plants, for example, are based on good practices.[25] But countries should be encouraged to examine their own situation as part of the environmental assessment process and to devise the best standards for local conditions. The new energy-environment sector work should additionally review emissions standards, the use of offsets within an airshed, and the opportunities for emissions trading.[26]

23. See, for instance, *Environmental Challenges of Fuel Use:* "Pollution Management in Focus" discussion, note 7. World Bank Environment Department, 1999.

24. RAINS-Asia (Regional Air Pollution Information and Simulation) is a project funded by the governments of the Netherlands, Norway, and Sweden through the World Bank and by the Asian Development Bank. The simulation model was developed by the International Institute of Applied Systems Analysis (IIASA) in collaboration with a number of scientific institutions in Asia and in the West.

25. These guidelines have been approved by the Operations Policy Committee and were made available to Bank staff in August 1998.

26. With regard to CO_2 emissions, an assessment of the implications for project design was carried out in *The Effect of a Shadow Price on Carbon Emissions in the Energy Portfolio of the World Bank: A Carbon Backcasting Study* (July 1998), which demonstrated how the recent energy portfolio of the

Better Analysis of Small, Non-Point Sources of Pollution

In most countries, roughly half the urban air pollution comes from small, widely distributed sources such as cars and cooking and heating fires. The *Pollution Prevention and Abatement Handbook* helps set emissions standards for large sources such as factories, but there is a need to develop a better strategy for non-point sources. Two-stroke engines are a particular and acute concern in many cities, for example; and the impact of car emissions can be reduced partly by phasing out lead from gasoline and by making other improvements in fuel type and quality. But any pollution strategy must also include an overall plan for the transport and urban sectors to reduce the negative impact on road transport. These sectors will be included within the Energy-Environment Reviews. The analysis developed in the recent UrbAir program will also be extended to cover many other cities and countries.[27]

Knowledge Management

Many country studies about energy and environment issues, such as the UrbAir reports, are currently available. Knowledge Managers within the Bank must make available the conclusions of these studies—future studies (and existing studies, to the extent possible) need to be made available on the Bank's internal Web site. The new sector networks are expected to help staff better share within the Bank their knowledge of environmental issues (Box 4.1).

Better Monitoring of Results

The absence of indicators of development outcomes in most of the energy-environment projects reviewed has made it virtually impossible to assess objectively the achievements of these projects. Typically, only

World Bank would have been impacted by the incorporation of a shadow value for carbon emissions. The study simulated what would have happened to project costs, types, and selection if the damages associated with global climate change had been integrated into energy lending through the use of a shadow value ranging from $5 to $40 per tonne of carbon emitted. The study also assessed low-carbon alternatives to the actual projects, thereby helping identify investment opportunities for greenhouse gas mitigation, for instance, through mechanisms such as Activities Implemented Jointly, or the proposed Prototype Carbon Fund.

27. In-depth reviews of existing and projected air quality in five metropolitan areas (Bombay/Mumbai, Metro Manila, Jakarta, the Kathmandu Valley, and Beijing) were undertaken in the UrbAir program. Final reports were published in October 1996.

engineering estimates are available, and these are based on the amount of hardware installed and the expected energy savings.

New project designs will include better monitoring and evaluation tools, such as the international protocol developed by USDOE for energy-efficiency projects. More effort will be needed during supervision to ensure that environmental benefits and energy savings are monitored and reported both by the borrower and by the Bank during implementation. Similarly, estimates will be made in project Environmental Assessments of levels of pollution expected from a project. These then can be monitored during the course of each project.

Strengthening Capabilities to Address Global Climate Change

Evaluate Climate Change Externalities as a Matter of Routine

Climate change externalities require different treatment from domestic pollution. They cannot be internalized in energy prices unless there is a demonstrated willingness to pay for them, for example through the GEF or Kyoto Protocol mechanisms. The first step toward integrating global environmental considerations into development will therefore be to ensure that quantification of greenhouse gas emissions from a new project is included in its environmental assessment. Bank Operational Policy 4.01 now requires that the greenhouse gas implications of all projects with potential global externalities be evaluated, and technical guidance has been issued to Bank staff. Environment staff now monitor progress toward full integration into the environmental assessment and safeguards process.

Climate change considerations should also be routinely integrated into Analytical and Advisory Activities and Energy-Environment Reviews. The results of these climate change "global overlays" will also help identify investment options for the Global Environment Facility, the Activities Implemented Jointly Program, and the Prototype Carbon Fund.

A Bigger Push for New Technologies

The biggest challenge is to create mechanisms to encourage the use of new technologies that reduce emissions of greenhouse gases. A crucial role in this for the Bank Group is to help open markets for commercial and near-commercial renewable energy and energy-efficiency technologies in its client countries. The Bank Group will accomplish this in the following ways:

■ Leveling the playing field through sector reform, and thus opening markets for renewable and energy-efficiency technologies.

■ Promoting renewable energy projects and energy efficiency as mainstream activities where they are cost-effective solutions to energy and environmental priorities.

■ Expanding support for the identification and preparation of renewable and energy-efficiency projects.

■ Assessing the risk/reward profile of emerging technologies and allocating risks to those best able to manage them through suitable project structures and financing mechanisms.[28]

Develop a Market for Carbon Emissions

The establishment of market-based mechanisms to help cut greenhouse emissions was agreed in principle at the climate change conference in Kyoto in December 1997, although details of implementation must still be worked out. The World Bank's Activities Implemented Jointly (AIJ) Program has served to demonstrate the market-based "joint implementation" mechanism, affirming the potential of a carbon offsets market. An efficient and equitable market in carbon emissions could mobilize substantial private sector resources, increase the development and spread of more energy-efficient technology to World Bank client countries, and enhance the energy and environment portfolio of the World Bank itself.

The Bank will take part in the critical development phase of this market. It will act as an intermediary, foster the establishment of a predictable market price for carbon offsets and credits, stimulate market

---

28. By using competitively awarded financing, whether concessional or at market rates, the WBG often can reduce the perceived risks of new technologies, thus stimulating investment in them. This has been demonstrated by projects such as the Poland Efficient Lighting Project (IFC/GEF), the China Efficient Industrial Boilers Project (Bank-GEF), and the Photovoltaic Market Transformation Initiative (IFC/GEF), and also by the recently approved IFC/GEF Renewable Energy and Energy Efficiency Fund.

growth through increasing participation and trading, help reduce trans-action costs, and encourage competition in the carbon-reduction invest-ment business. The Bank will also explore mechanisms to tap voluntary contributions through "green pricing" approaches and the development of "carbon neutral" product lines. The development of the Bank's new products in this area, including the Prototype Carbon Fund, will be sensi-tive to the pace of the UNFCCC process. It will take timing signals specif-ically from the outcome of the Fourth Conference of the Parties in Buenos Aires, Argentina, and subsequent progress in the international negotiations.

Introducing New Resources and Strategic Partnerships

Partnerships are direct, formal collaborations between the Bank and third parties, including other financial institutions such as the regional development banks, bilateral donors, foundations, nongovernmental or-ganizations, and the private sector. Specific partners will be chosen based on their skills, expertise, and resources for advancing the implementation of the energy-environment strategy. The evaluation of past partnerships shows that it will be important to share goals, expectations, and perform-ance standards. Wherever possible, partnerships will be used to leverage private sector-driven transactions and financial structures. Where a close partnership is not possible, effective coordination through donor round-tables and consultative meetings with UN agencies at the country level will help avoid duplication and confusion.

A New Partnership with the Global Environment Facility

Making the investments needed to reduce greenhouse gas emissions will often require expenditures beyond the level that is economically effi-cient for the borrowing country. But there are big opportunities for low-er-carbon investments that could attract new and additional resource transfers from the international community. We have initiated an ambi-tious new partnership with the GEF to increase overall funding directed at renewable energy opportunities. Employing instruments such as the APL model, long-term technology support mechanisms already employed successfully in OECD countries, and an increased reliance on in-country intermediaries to help identify and appraise projects, this partnership will seek to shift investments from one-off or demonstration projects to more effective long-term, country-based programmatic strategies. For private sector activities, the partnership will develop a more flexible decision authority to allow IFC to respond quickly to renewable energy investment opportunities.

Through the partnership, the GEF is expected to earmark approximately $150 million annually in grant and concessional financing for investment projects fitting the relevant GEF operational programs, to be allocated flexibly. More importantly, such partnership arrangements provide the basis for a sustained increase in financing commitments at manageable transaction costs and offer a comprehensive approach to renewable energy market and technology development. The initial framework of the partnership was recently endorsed by the GEF Council, and the Bank Group is now preparing pilot projects for review by the GEF later this year. Although intended initially for renewables, it is possible that this model could also be extended to use in other areas.

Bilateral Donors and Trust Funds

Much of the Analytical and Advisory Activities, project identification, and preparation have thus far been carried out with funds provided by bilateral donors. Greater financial support will be needed to implement the new strategy in general and the program of Energy-Environment Reviews in particular. The estimate of funds needed to carry out a program of 15 EER activities per year is in the order of an additional $1.5–2.0 million per year. The bulk of the financing of these activities will initially be borne by ESMAP, but in later years a greater share will be contributed from Bank administrative budgets.

More Public-Private Initiatives

The private sector is very active on the supply side in the energy sector and could become further involved on the demand side, particularly in terms of contributing energy-efficiency investments. One way to make these investments is through Energy Service Companies (ESCOs), which offer other firms services to help improve their energy efficiency. Because the institutional basis for ESCOs is lacking in most client countries, the Bank Group will begin the process of ESCO development through public-private sector partnerships within the framework of ESMAP, REEF, and direct Bank and IFC operations. More generally, the WBG will continue to expand its cooperation in the international energy industry through partnerships with groups like the Electric Power Research Institute (EPRI), the grouping of power companies from industrialized countries (the "E7"), and international oil companies.

Partnerships with Foundations

Studies are underway to assess the feasibility of the Solar Development Corporation (SDC). The SDC would help establish marketing and distribution channels for solar home systems, special photovoltaic units that provide household electricity without connection to a grid. Capitalization for the SDC would be provided through a partnership arrangement between the WBG, the GEF, several large philanthropic foundations, and other investors.

Bringing New Skills into the World Bank Group

One of the most inhibiting characteristics of operations involving the energy-environment nexus is that they are labor intensive. Part of this problem stems from the relative unfamiliarity of many World Bank staff and many borrowing countries with this sort of operation. As pressure mounts on Bank resources for lending operations, the interest of country and technical managers in undertaking more complex operations will inevitably decline unless this problem is explicitly recognized. At the minimum, more work is needed to assess how projects can be prepared and packaged to reduce their labor intensity, or at least their direct cost to World Bank budgets.

Implementing many of the initiatives in this strategy certainly will require "thinking outside the box." The Bank's new network structure across regions and the thematic sector groups should help in this respect, improving access to information and encouraging cross-fertilization of ideas. The thematic sector groups in particular are active in nurturing and supporting the development of projects and providing the necessary expertise and knowledge to bring those projects to realization.

The level and mix of staff skills needed to support the Bank's lending and non-lending services in the energy efficiency, rural and renewable energy, and energy and environment fields are now being reviewed by the sector boards for Environment and Energy, Mining, and Telecommunications. A reshaped program will require additional training of current staff through the World Bank Institute and the recruitment of new specialists.

IFC increased its commitment to sustainable energy in 1996 through formation of an Environmental Projects Unit within the Environment Division of the Technical and Environment Department. This unit is tasked with implementing IFC's GEF and Montreal Protocol projects, promoting new environmental investments for IFC's own account, and undertaking special initiatives such as working with the Bank on the Prototype Carbon Fund. In addition, IFC's Power Department has designated two senior staff members to coordinate renewable energy and energy-efficiency activities. These extra efforts are necessary because sustainable energy projects typically involve longer lead times and more complex appraisals than traditional energy investments; they are also smaller in size.

Implementing the Strategy: Trade-offs and Incentives

The implementation of this action plan calls for the deployment of resources to new activities. In an organization that is strictly bound by budget constraints, this implies that a smaller share of Bank resources must be allocated to other activities. Similarly, the focus on client demand will require active promotion to clients and Country Management Units.

Resources

A recent OED[29] report that analyzed the costs of preparing and appraising World Bank projects has concluded that costs increased significantly over the period FY 1989–96, and especially in the period FY 1994–96, when costs rose on average about 50 percent. In addition, three out of seven power projects that entered the lending program were dropped before Board approval. Cost increases and the large number of

29. World Bank. 1997. *The Effectiveness of the Bank's Appraisal Process: An OED Study.*

projects that were dropped can be attributed to the increasing complexity
of new projects, including those involving a broad range of development
objectives, such as environment, private sector development, and poverty
alleviation, and to the need to improve the quality of projects on entry
into the portfolio. Difficulties in using traditional lending instruments to
meet all objectives were also a factor.

The World Bank will make additional resources available to respond
to this challenge by (a) transferring operational costs upstream, (b) deepening sector reform and enhancing the environmental benefits of energy
lending through the use of new instruments, and (c) complementing
WBG resources with external funds.

■ **Transferring costs upstream.** The strategy emphasizes the introduction of Energy-Environment Reviews as a means of identifying the
issues and options to be addressed through lending and non-lending services. OED has found that where upstream sector work has been carried
out, project processing costs have been reduced and quality at entry has
been improved. This beneficial sector work will be done in the form of
EERs, transferring resources upstream from loan processing. As results
from the EERs are shared among several projects and sectors, they will
improve the quality at entry of both energy and non-energy projects.
Client participation in EERs will help to increase commitment to the
strategy during subsequent stages of the project cycle.

■ **Deepening sector reform and enhancing the environmental benefits of energy lending through the use of new instruments.** Initial experience with Adaptable Program Lending (APL) has been encouraging.
This instrument enables the Bank to respond flexibly to policy actions
undertaken by governments and to tailor financial resources to country
needs. APLs recognize the inherent uncertainty in changing client needs
and the tendency of governments to time their actions according to the
country's political cycle. The Bank will learn from the practical experience of expanding its energy-related environmental work and will likely
identify necessary modifications to the approach used. In addition, APLs
and LILs are good instruments for implementing the energy-environment strategy. There is therefore a likelihood that the overall share of traditional projects will decline.

■ **Prudent use of external funds.** Clearly, resources drawn from the GEF, ESMAP, and similar external sources have been vital in advancing the environmental agenda in energy. Indeed, much of the pioneering work on emerging good practice in this area has been funded externally. While these partnerships must continue, the World Bank's operational budget will probably take up the main burden of mainstreaming as soon as the pilot phases of the new approaches end. Without this shift, the new agenda would remain marginal. External funds will be deployed at the cutting edge as seed resources for new projects, and upstream for preparatory sector work, complementing the operational budget in EERs. This prudent use of partnership funds will ensure that the driving force of the energy-environment agenda firmly resides in Bank operations, and that the external resources are applied as "venture capital" to explore and demonstrate promising good practice.

For its part, IFC is responding to the increasing role of the private sector in the provision of energy services in many parts of the developing world. In addition to its investments in the traditional energy sector, IFC will support and encourage new investments in the alternative energy sector. As discussed above, it is dedicating both increased budget and manpower to develop renewable energy and energy-efficiency projects, and expects to see growth in these types of investments. IFC will continue to make prudent use of concessional funding from sources such as the GEF in order to provide incremental assistance for near-commercial projects that aim to accelerate market acceptance of alternative energy technologies and services.

Incentives

The strategic priorities will be integrated in a demand-driven operational environment by (a) aggressive promotion of good practice in energy/environment operations, (b) use of new lending instruments and non-lending services, and (c) the judicious use of external funds as incentives to enter into new operations.

■ **Promotion of good practice.** A first exploratory round of consultations with the Country Directors responsible for key clients with the potential demand for energy/environment assistance yielded significant interest in upstream sector work to identify areas of possible intervention. These early leads will be followed up vigorously, and the range of client consultations will be expanded. Emerging good practice in implementing energy efficiency, renewable and clean fossil energy applications, and integrating environmental regulation and standards in sector reform will be promoted widely as viable operational alternatives. The relevant Thematic Groups in the Private Sector and Infrastructure (PSI) and Environmentally and Socially Sustainable Development (ESSD) networks will play a key role in advocacy and demonstration.

■ **New instruments.** The use of new lending instruments such as APLs and LILs will enable learning-by-doing, and will thus reduce the perceived risks of operations designed around the new agenda (for example, sector reform in India and energy efficiency in Brazil). This will help to alleviate the natural caution that meets a new type of operations that lacks a proven track record. Similarly, increased use will be made of non-lending services to transform and enable markets to handle environmentally benign energy. Often, this is the most effective way to bring the environmental agenda into energy with relatively modest resources; regulatory and standards reform can be handled this way, for example.

■ **Funding incentives.** A careful application of external funds as upstream project identification vehicles, or as sector work that sets the scene for operations, will act as an incentive to generate demand. Partnership funds such as the GEF and bilateral project development funds have already proven to be powerful incentives for clients and Country Management Units to focus on the new agenda and to consider a new type of operations. Upstream sector work funded by ESMAP and project identification by the Asia Alternative Energy Program and by the nonconventional and traditional energy funds for Africa have laid the groundwork for environmentally driven operations under preparation or in the portfolio. A key need is to make sure that incentives to use external funds reinforce the objective of meeting clients' highest priorities.

Accurate monitoring of progress in implementing this strategy will be done using indicators that focus on the outcomes of interventions rather than on their inputs—how much World Bank Group activities contribute to our clients' comprehensive development, rather than how much we invest in our activities. Outcome indicators are linked to programs of action and forecasts of short-term outputs based on what clients are able to implement. This approach leads to strategic work programs designed to achieve environmental results that actually improve the lives of the poor. A summary of the indicators is set out on pages 12 through 17, and more detailed plans for each Bank Region are described in Annexes 5 through 10. Progress will be regularly assessed against these indicators to provide feedback to staff and managers and to enable them to fine-tune the strategy. The World Bank Group will inform the Board of progress after one year, and will report formally in two years' time.

REFERENCE MATERIAL

Annex 1: Environmental Impacts of Energy Use

Sector/Subsector	Activities	Impacts/Issues
Coal/Lignite	Mining, processing, and transport	Land use and degradation, air quality, and social impacts. Effect of road, rail, and maritime transport systems on land use, waterways, and coastal zones
Oil and Gas	Extraction, refining, and transport	Land degradation, spills/leaks, flaring having impacts on land use, and airshed. Effects of/on pipeline, road, and rail on land use, waterways, and coastal zones
Fossil-Based Electric Power	Fuel conversion, transport, and distribution	Airshed, land use, waterways, and coastal zones with knock-on effects on health, productivity, and deterioration of infrastructure. Significant increases in emissions normally result from system expansion
Industrial Heat Raising	Fuel conversion and distribution	Similar to fossil-based electric power (emissions, air quality)
Transport	Fuel conversion	Use of petroleum products (increasing use of alternative fuels such as LPG, CNG, and biofuels)
Domestic	Fuel conversion	Use of primary and secondary energy. Indoor air quality
Traditional Fuels	Fuel conversion	Production and use of biomass. Impacts on agriculture ecosystems, woodland, and forests. Land degradation. Indoor air environment and health impacts for end users. Social dimensions
Fuel Mix	Fuel switching	Dependence on imported fuels, diversification of fuel sources. Changes to processing and distribution systems. Generally positive effects on household, local, regional, and global environments. Possible social impacts
Energy Efficiency	Supply-and demand-side efficiency measures	Improved financial/economic performance. Generally positive impacts by avoiding commissioning of new sites/facilities
Renewables	Energy production	Use of "greenfield" sites and impacts on airshed, land use, groundwater, and noise. Accelerated replacement or decommissioning of old systems and sites
Cleaner Fossil Fuel Systems	Fuel conversion	Airshed, land use, and accelerated replacement or decommissioning of old systems and sites
Reform		Effects on pricing, taxation, and levies with knock-on effects in physical environment

Annex 2: The Costs of Local and Regional Air Pollution and Global Climate Change

Introduction

The purpose of this annex is to draw attention to the damage—local, regional, and global—that results from energy use. Local damage includes adverse health effects from indoor air pollution as well as the effects on human health and ecosystems of outdoor air pollution. Regional air pollution encompasses ground-level ozone, which can damage human health, crops, and forests, and acid deposition, which can degrade ecosystems. The global damage associated with energy use includes the effects of global climate change, such as sea-level rise, catastrophic weather events, damage to agriculture and human health, water shortages, and the amenity effects of changes in temperature and precipitation.

Wherever possible, the environmental costs of fuel use are expressed in quantitative terms. Obviously, a quantitative assessment cannot provide more than a rough indication of the likely order of magnitude of environmental costs. Analytical difficulties, a lack of data, our still poor understanding of some of the phenomena, and ethical questions related to the comparison and aggregation of results all make the calculation of exact numerical estimates elusive. However, even a rough assessment of the environmental costs of energy use can provide useful policy insights. Two insights in particular stand out.

First, the damage associated with energy consumption is large. Conservatively estimated, the present discounted value of local health damage from fuel consumption over the next 100 years (compared to ambient PM_{10} levels of 20 micrograms per cubic meter) are in the range of $8 to $19 trillion (in 1995 U.S. dollars), depending on the discount rate used. This is the dollar cost of 3.5 million excess deaths per year, plus the cost of respiratory illnesses resulting from air pollution. The present value of global damage associated with fuel use—also conservatively estimated—is approximately $2 to $12 trillion (1995 USD), depending on the choice of discount rate.

Second, developing countries bear a disproportionate share of this burden. Of the 3.5 million deaths associated with local air pollution each year, 3.25 million occur in developing countries. Developing countries bear 86 percent of the total dollar burden of local air pollution—equivalent to 12.8 percent of GDP annually in India and 13.8 percent of GDP annually in Sub-Saharan Africa averaged over the period 2001 to 2020. This occurs despite per capita losses from illness and premature death being much lower in developing countries than in Established Market Economies (EMEs). The damage caused by climate change (assuming a $2.5°C$ increase in mean global temperature) is conservatively estimated to reach 4–5 percent of GDP annually in India and Sub-Saharan Africa by 2100. Many developing countries will need to implement long-term measures to adapt to the consequences of climate change.

The size of this damage suggests that attention should be paid to the external costs of energy consumption. Policies will need to be considered that will reduce local air pollution and the emission of greenhouse gases (GHGs). A comparison of the total local damage and total global damage from energy use (such as the one provided here) cannot by itself be used to set priorities for air pollution control, however. Which policies should be undertaken depends not on total damage, but on the net benefits of the policies—i.e., on the benefits of pollution reduction minus the costs of abatement. It also depends on who bears the costs of abatement and on who enjoys the benefits.

The numbers presented on the following pages suggest that the damage caused by local air pollution in developing countries is high, and that countries should consider policies to reduce that damage. Bank studies that have examined the benefits and costs of policies to reduce local air pollution have demonstrated that, for some policies, the net benefits are large.[30] Which policies pass the benefit-cost test will vary from country to country, depending on per capita income, prices, and, in the case of outdoor air pollution, on population density.

While some policies to reduce local air pollution may have little impact on GHG emissions (for example, removing lead from gasoline or sulfur from diesel fuel), many other policies that reduce local air pollution will also reduce GHG emissions (for example, substituting natural gas for coal in home heating and electricity generation, or increasing the use of renewable energy technology). Which policies should be given priority depends on their net benefits. To the extent that industrialized countries are willing to defray part of the cost of measures that reduce emissions of GHGs in developing countries—for example, through the Global Environment Facility or the emerging market for carbon credits[31]—the domestic balance of net benefits will shift in favor of actions that lead to lower emissions of both local and global air pollutants. . . .

30. See, for example, the following: World Bank. 1997. *Can the Environment Wait?* World Bank. 1994. "The Net Benefits of an Air Pollution Control Scenario for Santiago." In *Managing Environmental Problems in Chile: Economic Analysis of Selected Issues.* Jian Xie, Jitendra J. Shah, and Carter J. Brandon. 1998. "Fighting Urban Transport Air Pollution for Local and Global Good: The Case of Two-Stroke Engine Three-Wheelers in Delhi." World Bank: Mimeo.

31. The Bank's Prototype Carbon Fund, which invests in emission reduction credits for CO_2, provides a method of financing GHG reductions in developing countries.

Methods

This annex summarizes the findings of a recent study that set out to estimate the overall cost of the environmental damage associated with energy use[32] and its composition.[33] Drawing upon a wide range of sources, the study provides a breakdown of costs that distinguishes between damage to human health and other types of damage, and also between different types of exposure considered in terms of the relationship between the fuel user and those who are affected by the damage. The four exposure categories are:

1. Direct exposure to indoor air pollution caused by the use of dirty fuels for cooking or heating in poorly ventilated houses.

2. Local exposure to emissions from low-level sources such as stoves, boilers, and vehicles that affect people living within a small distance—usually 1–5 km, but up to 20 km—of the original source.

3. Regional exposure to pollutants that may travel or be transformed in the atmosphere over a distance of 50–500 km from the original source, as in the case of acid depositions.

4. Global exposure to pollutants for which the total volume of emissions around the world is the key determinant of the damage caused, such as the emission of GHGs.

Categories 1 and 2, which together cover the cases in which the effects of pollution are felt almost entirely within the areas covered by national or local governments, are jointly referred to as "local pollution" in the study and in this annex. Categories 3 and 4 can only be effectively addressed by collective action involving groups of countries in a region or the world community as a whole. They are referred to as "regional" and "global" pollution, respectively.

The methods used in the study for valuing the damage caused by air pollution are consistent with the methodology applied for country studies carried out on behalf of the U.S. Environmental Protection Agency (US-EPA) and the European Commission (EC). There are a number of important technical issues about the application and interpretation of some of the assumptions. Standard methods of valuation are employed to consider marginal changes in exposure and risk. On the other hand, any estimate of the total cost of environmental damage

32. Emissions caused by energy use may be only one of a number of sources of the environmental damage analyzed in the study. For example, rice cultivation, livestock production, and changes in forest cover all contribute to the accumulation of carbon dioxide in the atmosphere. The paper focuses on environmental damage for which energy use is generally the most important contributory factor, but the methods tend to overestimate the total damage caused by energy use in cases where other factors are significant.

33. A detailed description of this study is given in Lvovsky, K., G. Hughes, D. Maddison, B. Ostro, and D. Pearce. (Forthcoming.) *Air Pollution and the Social Costs of Fuels.* Washington, DC: World Bank Technical Paper.

refers to the non-marginal changes involved in moving from the current situation to a hypothetical reference case in which the exposure to air pollution meets
a target set of ambient standards. To address these concerns and other uncer-
tainties, the assumptions used in this study to value the costs of local air pollution were explicitly chosen to underestimate these costs. Applying the less con-
servative assumptions adopted in the US-EPA and EC studies would increase the
estimates of the damage caused by local air pollution by 50–100 percent, but
would not change the broad pattern of results.

Comparing the costs of local, regional, and global air pollution for the world
as a whole and for specific regions raises two major concerns that were not as
important in previous studies:

■ How should differences in income across countries be incorporated in the
analysis? From one perspective it may be argued that the premature death of a
person aged 50, or a year of disability caused by ill-health, should be valued
equally, no matter where in the world the person or persons affected live. Equally, it is clear that willingness to pay to reduce the risk of premature mortality and
health is a function of current and/or future income, so that valuations should
take account of differences in income across countries. Reasonable cases can be
made for several different sets of assumptions. Which is most appropriate depends largely upon an ethical judgment about the manner in which differences
in income or living standards should be incorporated in the analysis. The primary estimates reported below assume that valuations are based on each country's current and predicted future GDP per person. If, instead, a world average
level of GDP per person had been used for all valuations, the total cost of damage caused by air pollution would have been more than 10 times higher, with the
largest increase for local air pollution, because the largest part of the damage is
suffered by low-income countries.[34]

■ How should the analysis allow for differences in the timing of the costs of
different types of exposure? The effects of the accumulation of carbon dioxide
and other GHGs in the atmosphere will be felt over decades and centuries, while
the costs of local and regional air pollution are manifest in deaths, ill-health, and
other consequences that affect the welfare of people today and in the near future. The answer to this question is bound up with the treatment of differences
in income, since they revolve around attitudes to inequality both across genera-

---

34. Focusing either on specific physical measures of damage such as the burden of disease measured in DALYs or on valuations that are based on a world average level of income would direct attention to measures that reduce gross inequalities between countries in life expectancy, quality of
life, and absolute poverty. Rapid economic growth is usually a necessary condition for achieving
these goals. In this context, energy-environment policies would emphasize improvements in the
macro-regulatory framework that would ensure investment and other resources are utilized in the
most efficient manner.

tions and across populations at a single point in time. The "equal weights" assumption implies the use of world average GDP per person for valuation, in combination with a low discount rate—assumed to be 3 percent per year—for aggregating effects over time. The "current weights" assumption implies the use of current/future national GDP per person combined with a discount rate that falls gradually over time, on the assumption that economic growth will lead to a convergence in income levels. The primary estimates reported here assume that the discount rate applied to costs declines from 8 percent per year to 3 percent per year on a straight-line basis over 100 years.

The analysis of the costs of direct, local, and regional exposure requires projections of emissions and exposures over the next 100 years. Such projections entail heroic assumptions and are, of necessity, highly uncertain. Costs are based, as far as possible, on detailed estimates and projections of emissions and exposure on a country-by-country basis. Some key parameters—notably rates of population growth, urbanization, and GNP per person—are only available on a regional basis for the next 100 years. Thus, the estimates reflect the situation in 1995 with respect to air pollution, composition of energy use, and income, combined with regional economic and demographic trends that determine the evolution of exposure and costs over the full time horizon.

The costs of global climate change are based upon a recent study by Nordhaus and Boyer.[35] Their estimates were obtained from a quasi-sectoral basis covering (a) agriculture, (b) sea-level rise and coastal vulnerability, (c) other vulnerable market sectors, (d) health, (e) non-market amenity impacts, (f) human settlements and ecosystems, and (g) catastrophic impacts. The estimates of health costs were adjusted for consistency with the rest of the analysis. The effects of an increase in average temperature of 2.5°C, which is expected to occur shortly after the year 2100 in their model, are estimated for 13 regions. The population-weighted global cost is estimated at 1.9 percent of projected GNP each year, of which 55 percent is accounted for by willingness to pay to avoid or mitigate catastrophic impacts. For this analysis it was assumed that the 2.5°C costs apply in the year 2100 and that costs for intermediate years may be obtained by linear interpolation over the period 1995–2100. It must be emphasized that these point estimates are highly uncertain. While this is true to some extent of the health costs of air pollution, there is greater uncertainty regarding the effects of climate change.

35. See Chapter 4 of: Nordhaus, W.D. and J. Boyer. (Forthcoming.) *Roll the DICE Again: Economic Models of Global Warming.* Cambridge, Mass.: MIT Press.

An earlier, and somewhat more disaggregated, version of the same work was reported in Nordhaus, W.D. 1998. "New Estimates of the Economic Impacts of Climate Change." Yale University: Department of Economics.

There are a variety of other estimates of the costs of global climate change.[36] While these estimates use more accurate representations of climate dynamics than the assessment used for this paper, it was not possible to use them here, either because the results are presented solely as present values using fixed discount rates, or because they are not sufficiently disaggregated—by region/country and type of damage—to permit adjustment for consistency with the assumptions used in the rest of the analysis. Quantitative estimates of the impact of climate change are still highly uncertain. Estimates of market impacts in developed countries have tended to fall as investigators have improved their analysis. On the other hand, there is growing concern about non-market damage (e.g., on ecosystems) and the impact on developing countries. The Nordhaus and Boyer estimates are often considered as falling at the low end of the range.[37]

Results

The largest component of the costs of air pollution is the result of premature mortality and ill-health caused by direct and local exposure to high levels of pollutants. Table A2.1 summarizes the projections of the burden of disease caused by air pollution over the period 2001–2020 under a "business-as-usual" scenario as annual averages for the 20-year period. Since a substantial proportion of the impact takes the form of ill-health of differing degrees of severity rather than premature deaths, the estimates of the total burden of disease are expressed in terms of disability-adjusted life years (DALYs), which are calculated using a standard approach adopted by the World Health Organization (WHO).[38]

The projections indicate that about 3.5 million people will die prematurely each year over the next 20 years as a result of indoor and outdoor air pollution. In Sub-Saharan Africa and South Asia (excluding India), most of these deaths will be the result of indoor air pollution. In India, indoor and outdoor air pollution each account for about one-half of the total number of premature deaths; in China, the former socialist economies, and Latin America, outdoor air pollu-

36. The most careful recent study is: Tol, R. 1999. "The Marginal Damage Costs of Greenhouse Gas Emissions." *Energy Journal*, Vol. 20 (1), 1999, pp. 61-81.
For an overview, see: Pearce, D.W., et al. 1996. "The Social Costs of Climate Change: Greenhouse Damage and the Benefit of Control." In IPCC: *Climate Change 1995. Economic and Social Dimensions of Climate Change. Contributions of Working Group III to the Second Assessment Report of the Intergovernmental Panel on Climate Change.* Cambridge, U.K.: Cambridge University Press.
37. A sensitivity test was carried out by combining the results of using estimates of global damage that are 50 percent higher than the Nordhaus and Boyer estimates—i.e., that are roughly equivalent to the figures up to 2030 derived in the study by Tol cited above—with the estimates of damage caused by direct, local, and regional exposure that are consistent with the EPA and EU estimates discussed above. This generates ratios of local to global damage that are larger than those shown below.
38. See Murray, C. J. and A. D. Lopez, eds. 1996. *The Global Burden of Disease.* Cambridge, Mass: Harvard University Press.

Table A2.1 Premature Mortality and Burden of Disease Due to Air Pollution, by Region
(Projected annual averages for 2001–20)

Region	Premature Deaths (thousands per annum)			Burden of Disease (DALYs per annum, millions)		
	Direct	Local	Total	Direct	Local	Total
China	150	590	740	4.5	14.0	18.5
East Asia and the Pacific	100	150	250	3.5	3.8	7.3
Established Market Economies (EMEs)	—	20	20	—	0.5	0.5
Former Socialist Economies	10	200	210	0.2	3.8	4.0
India	490	460	950	17.0	10.1	27.1
Latin America and the Caribbean	10	130	140	0.3	3.7	4.0
Middle East and North Africa	70	90	160	2.4	2.5	4.9
South Asia	220	120	340	7.6	2.6	10.2
Sub-Saharan Africa	530	60	590	18.1	1.2	19.3
World	1,570	1,810	3,480	53.4	42.2	95.6

tion is the major issue. The total burden of disease amounts to almost 100 million DALYs per year, or the equivalent of a loss of more than one year of life over the lifespan of the average individual. For India and Africa, the burden of ill-health caused by air pollution is equivalent to a loss of life of more than 1.5 years.

Estimates of the present value of the damage associated with air pollution depend critically on the assumptions used to monetize benefits and to discount them to the present. When, in addition to the assumptions above, it is assumed that the value of avoiding a DALY is five times per capita income, the present value of the damage caused by air pollution from 2001 to 2100 is estimated to amount to $11.9 trillion at 1995 prices. Of this total, $7.6 trillion (64 percent) is linked to direct and local exposure—i.e., local pollution—while climate change accounts for $2.6 trillion (21 percent). If the value of avoiding a DALY is estimated to be twice per capita income, the present value of local health costs equals about $3 trillion and the value of global damages about $2 trillion. If, on the other hand, the value of a DALY were consistent with the assumptions used by the US-EPA and the EC in evaluating the benefits of local and regional air pollution policies, the present value of local health costs would equal about $18 trillion and the value of global damages about $3.6 trillion. As noted in the introduction to this paper, variations in the discount rate can cause the present value of damages from climate change to vary by a factor of six. The share of local pollution, however, exceeds that of climate change in all of the scenarios examined in the study.

The aggregate costs of air pollution for the world hide large variations between countries and regions in the relative importance of different types of exposure to the overall total damage. Table A2.2 breaks down the costs by region, using a classification that closely corresponds to that used for the World Bank/WHO study of the global burden of disease in 1990, cited above. Here, the WHO region called Other Asia and Islands, and which excludes China and India, has been split between the separate East Asia and Pacific and South Asia regions.

The figures in the table illustrate why the concerns of different countries or groups of countries may diverge so widely. Contrast, for example, the EMEs and the rest of the world. For the former, the damage caused by local, regional, and global exposure is almost equally important, while the damage caused by direct exposure to air pollution is negligible. For the low- and middle-income countries as a group, the value of damage caused by regional exposure is small by comparison with the same damage for the EME region and, more importantly, relative to the cost of damage caused by direct and local exposure. As noted above, the relative importance of aggregate local and regional damages has limited bearing on abatement priorities, which depend on the net benefits of individual policies.

Conclusions

The burning of fuels has large environmental consequences, and developing countries are particularly badly affected. Growth in the use of energy services is nonetheless fundamental to the development prospects and quality of life of many of the same people who are affected by the environmental damage that those energy services cause. Any exercise in policy formulation must, therefore,

Table A2.2 Present Value of Damage Due to Air Pollution, by Region

Region	Present Value of Damage Due to Air Pollution (1995 USD, billions[39])				
	Direct	Local	Regional	Global	Total
China	245	1,330	121	22	1,720
East Asia and the Pacific	200	512	62	286	1,060
Established Market Economies (EMEs)	0	1,030	1,360	1,150	3,540
Former Socialist Economies	0	641	73	27	742
India	481	605	23	252	1,360
Latin America and the Caribbean	39	1,070	56	357	1,530
Middle East and North Africa	100	483	43	168	794
South Asia	247	164	3	50	464
Sub-Saharan Africa	389	71	21	184	665
World (excluding EMEs)	1,700	4,880	402	1,350	8,330
World	1,700	5,910	1,760	2,500	11,900

39. All figures have been rounded to three significant digits. Rows and columns may not add to totals shown due to rounding.

strike a balance between the costs and benefits of increased energy use, and must take account of the opportunities to reduce the impact of energy use on the environment at low or moderate cost. There are few universal guidelines—even these will change in the face of new technologies and evolving priorities. Nonetheless, it is important to ensure that policies are grounded in a realistic quantitative appraisal of the likely costs and/or benefits of alternative courses of action.

The fact that direct and local damage is such a large component of the total damage caused by fuel use suggests that these factors will dominate the estimated value of domestic benefits from specific air pollution control policies. Low-cost policies that control local air pollution and that address climate change may yet emerge, however. Whether or not the policies that yield the greatest net benefits to developing countries are ones that focus exclusively on local air pollution control is an empirical question. To answer it countries must look at the costs as well as the benefits of pollution control.

Annex 3: Environmental Strategies Record

Category	Portfolio Analysis				Current Non-Lending: TA/Policy Work	Key Barriers, Constraints, and Issues Affecting Further Work	Remarks
	Actual Lending (three-year total, FY 1996–98)		Pipeline (three-year total, FY 1999–2001)				
	Number of Projects	Funding: All Bank Sources ($ millions)	Number of Projects	Funding: All Bank Sources ($ millions)			
End-Use Efficiency	13	571	11	280	Energy-efficiency technology transfer, ESCOs, DSM, and building thermal performance. Disseminating best practices and introducing market-based energy conservation mechanisms	■ continued low oil price ■ market distortions ■ high transanction costs ■ financing	Pipeline needs to be further developed. Success of pilot projects now under way will affect pace of pipeline development
District Heating	4	261	7	471	Metering, billing reform, and distribution network efficiency. Reducing system losses and removing barriers to user-oriented energy efficiency improvement	■ weak institution financing	Began only after economic transitions were under way. Large potential in Eastern Europe and the CIS for rehabilitation projects with private sector participation. Large potential in China in environment projects
Supply-Side Efficiency and T&D Loss Reduction	7	410	10	536	Electricity T&D loss reduction, power plant rehabilitation, and natural gas transmission loss reduction. Reducing serious losses of energy in the supply chain, improving energy system O&M efficiency, and strengthening energy system management	■ Bank is less involved with increasing privatization, resulting in reduced leverage	Half of power sector projects have components or conditionality to reduce losses. Need to increase focus on rehabilitation and environmental improvements in existing power plants. Need balance between transmission, distribution, and generation

(Continues)

	Portfolio Analysis						
	Actual Lending *(three-year total, FY 1996–98)*		Pipeline *(three-year total, FY 1999–2001)*				
Category	Number of Projects	Funding: All Bank Sources *($ millions)*	Number of Projects	Funding: All Bank Sources *($ millions)*	Current Non-Lending: TA/Policy Work	Key Barriers, Constraints, and Issues Affecting Further Work	Remarks
Environmentally Innovative Energy	6	201	6	100	Controlling air pollution from energy use through rehabilitation and conversion. Piloting innovative schemes and building air quality management capacity	▪ institutional barriers in utilities incentives	Focus to include pollution mitigation and strengthened capacity of environmental agencies
Renewable Energy (On- and Off-Grid)	11	274	23	902	Capacity building for implementation. Project identification and mobilization of donor funding	▪ continued low oil price ▪ carbon trading mechanisms in place ▪ transition away from long-term contracts ▪ Bank lending emphasis on rural development and environment	Strong pipeline, achievable pace depends on implementation and macro climate. For geothermal, shift from power generation heat production and co-generation. Need to interest IPPS
Oil and Gas (Environmentally Innovative Only)	4	174	4	86	Reform/restructuring of gas industry and further commercialization. Gas flaring reduction. Improvements to product specification/quality	▪ restructuring of oil industry due to low oil price ▪ expanded oil and gas trade	Need to include policy, pricing, and privatization issues in country dialogue
Power Sector Reform	31	N/A	94	N/A			Included in nearly all power sector projects. Privatization support in 20 projects in 16 countries
Coal Sector Reform	4	N/A	5	N/A			Emphasis on sector reform in production
TOTAL	80	1,891	160	2,375			

NOTE: Bank includes IBRD, IDA, and IFC operations. Reform elements are included in most projects. Lending has been omitted to avoid double counting.

Annex 4: Existing World Bank/IFC/GEF Renewable Energy Initiatives

Table 1 Existing World Bank/IFC/GEF Country-Specific Renewable Energy Initiatives

Coverage	Project	Technologies	Status (Board Approval)	Bank	GEF	Other	Total	Description	GEF Element
Cape Verde	Energy and Water Sector Reform	PV, wind	Proposed FY 1999	18	5	25	48	Investment in wind farms and PV market	Wind and PV cost buy-down
China	Renewable Energy Development Project	PV, wind	Proposed FY 1999	100	35	273	408	Investments in wind farm and PV market development	PV cost buy-down and TA
India	Solar Thermal	Solar thermal	Proposed FY 1998		49	196	245	Demonstrate and promote solar thermal power generation	Equipment buy-down and TA
India	Renewable Energy II	Small hydropower, energy efficiency	Proposed FY 1998	150		113	263	Credit line for 200 MW of small hydropower investments	None for renewable component
India	Renewable Resources Development Project	PV, small hydropower, wind	FY 1993	115	26	139	280	Credit line for PV and wind, and TA for private sector market development	Reduced rate on loan to PV and wind only
Indonesia	Second Rural Electrification Project	Mini-hydro, mini-geothermal	FY 1995	22.5		13	35.5	Credit line and TA to strengthen institutional RE planning capacity; mini-hydro and geothermal testing investments	None
Indonesia	Solar Home Systems Project	PV	FY 1997	20	24.3	73.8	118.1	Catalyze investments by providing credit, equipment buy-down, and TA	Grant for equipment buy-down and TA
Indonesia	Eastern Indonesia Renewable Energy Development Project	Small hydropower, small geothermal	Hold	63.8	3	31.4	98.2	Catalyze investments, and mainstream RE planning for remote areas.	Grant for TA only
Lao PDR	Southern Province Rural Electrification Project	PV, small hydropower	FY 1998	1	0.74	0.1	1.84	Pilot off-grid electrification for future expanded project	Grant for TA and equipment buy-down
Pakistan	Lahore Waste-To-Energy Project	Grid-connected biomass (MSW)	Hold	2	10	tbd	at least 12	Support for new power generation facility from landfill gas; and TA	Grant covers incremental cost compared to standard landfill
Philippines	Renewable Energy Applications on Island Grids	mini-hydropower, wind, geothermal	Feasibility study			3.6	3.6	Demonstrate commercial RE applications for island grids	Under AIJ program with Swiss Government
Sri Lanka	Energy Services Delivery Project	PV, mini-hydropower, wind	FY 1997	24.2	5.9	25.2	55.3	Credit for SHS/hydro, wind pilot project, and TA	Grant for PV, village hydropower, and TA for wind

Table 2 Existing World Bank/IFC/GEF Worldwide Renewable Energy Initiatives

Coverage	Project	Technologies	Status (Board Approval)	Funding ($ millions)				Description	GEF Element
				Bank	GEF	Other	Total		
Worldwide	IFC-GEF Renewable Energy and Energy Efficiency Fund (REEF)	Renewable energy	IFC approval: FY 1997 GEF approval: FY 1998	25–35	20–30	75–165	120–230	Catalyze investment by taking debt and equity position in small (<50 MW) projects	Concessional cofinancing for smaller/ riskier projects through grant, loan, equity, and guarantee. Also fund management
India, Kenya, Morocco	IFC-GEF PV Market Transformation Initiative (PVMTI)	PV	GEF approval: FY 1998	tbd	30	45–70	75–100	Accelerate development of private sector companies providing PV goods and services	GEF cofinancing through grant, loan, equity, and guarantee. Concessionality as proposed by competing sponsors
Worldwide	Solar Development Corporation (SDC) IBRD/IFC/GEF Foundations	PV	IBRD initial approval: FY 1999 IFC approval: FY 1999 GEF approval: FY 2000 (projected)	12.5	10	27.5	50	Accelerate growth of off-grid PV market worldwide	Proposed business advisory compo- nent of $18M (grant), and some concessional investment in $32M finance window
Worldwide	IFC-GEF Project for Small and Medium-Sized Enterprises	Renewable energy, esp. PV	GEF approvals in FY 1996 and FY 1997	tbd	21	19–29	40–50	Catalyze projects by funding incremental costs through financial intermediaries, including NGOs	Loans to FIs at 2.5 percent, 10 years for onlending to SME on negotiat- ed terms. FIs keep portion of capital recovered

NOTES: tbd: to be determined; FI: financial intermediary; MSW: municipal solid waste; SHS: solar home system; SME: small and medium-sized enterprises.

Annex 5: Monitorable Progress Indicators for the Africa Region

Strategic Objectives	Outcomes	Actions Needed	FY 2000–02 Outputs
More sustainable production of traditional fuels plus improved efficiency of utilization. (Use traditional fuels more efficiently)	Sustainable harvesting More efficient charcoal production More efficient end use	Reform to achieve participatory management of resources, with significant role for local communities Knowledge: Improved resource husbandry and charcoal manufacture, plus better stoves Financing: Education of NGOs/communities; tools, stove manufacture	Additional countries participating in the RPTES program Three additional traditional fuels activities/project components Significant increase in acreage of forest under sustainable management
Improve basic access to commercial energy among rural and periurban populations. (Promote substitution away from traditional fuels)	Ease reliance on traditional fuels Growth opportunities for rural and peri-urban population Promote market for new energy technologies	Develop new approach focusing on local private sector and communities in liberalized sector structure Knowledge: Devise distributed technical solutions, with emphasis on new energy technologies Financing: Sources of equity and debt for small local schemes; guarantees to bring in local commercial banks	Design of sustainable institutional approach Three full-scale schemes launched Wide dissemination, with at least one country proceeding independently of the Bank
Improve technical and financial efficiency of energy utilities. (Cut supply losses in utility sector)	Private participation in utilities Competition in bulk supplies Higher technical efficiencies and better financial performance	Reform of utility sectors: Unbundling, public/private partnerships, privatization, modern utility regulation Knowledge: Cross-fertilizing reform success. Advisory support for planning privatizations and modern regulation Financing: Support for privatizations and project finance transactions (with IFC) as and when needed	Two, possibly three, privatizations completed Financial closing of two landmark private transactions Other reform programs underway with stronger financial performance among utilities
Encourage regional integration of energy infrastructure. (Cut losses through scale economies and avoiding duplication of supplies)	Regional integration of power grids New supply projects planned on regional rather than national basis Wider, more competitive energy markets	Promote fuller understanding of environmental, economic, and strategic advantages of integrating supply Knowledge: Exposure of countries to best practices in grid organization, regulation, trading arrangements Financing: Missing transport links; key regional infrastructure	Support the establishment of two regional energy systems One regional system operational Support for one, possibly two, important integrating projects
Reduce gas flaring. (Cut carbon emissions)	Reduced flaring of gas, lower emissions Economic benefits from utilization of clean, low-cost resource Improved integration of energy producers with neighboring countries	Improved dialogue with producing countries Policy reform to curb flaring Infrastructure to transport and utilize gas	Action by at least one country to curb flaring Support for regional pipeline in West Africa to collect flare gas Support for at least one private project to utilize flare gas

Annex 6: Monitorable Progress Indicators for the East Asia and Pacific Region

Strategic Objectives	Outcomes	Actions Needed	FY 2000–02 Outputs
Improve access to modern energy in rural areas to alleviate poverty and reduce health impacts of traditional fuel use	Improve access by rural households to cleaner commercial energy Increase share of cleaner fuels by 10 percent in rural sector of at least one country by 2005 and by 20 percent in at least two countries by 2010	**Short and medium terms (FY 2000–03)** Promote access to and use of electricity and other clean fuels in rural areas in at least three client countries Develop renewable energy for off-grid rural electrification as well as grid-connected power sector Introduce pilots for renewable energy development across non-energy sectors (industry, rural, urban, education, and health) in at least three client countries **Long term (FY 2004–08)** Integrate environmental considerations into all Bank rural energy operations Mainstream renewable energy in all rural energy projects in client countries	Four rural energy projects (Cambodia, Lao PDR, the Philippines, and Vietnam) will promote access to cleaner fuels, including renewables Indonesia: A rural electrification project emphasizing renewable energy may be reintroduced in FY 2002 Vietnam: Rural Energy Project for FY 2000, which would have a renewable energy component Sector work: A rural electrification policy note for the Philippines in FY 2000. A renewable energy strategy will be prepared for Vietnam Non-energy sectors: Capacity building and pilot projects for using renewable energy in the education and health sectors of Vietnam and China
Improve air quality in urban areas	Improve air quality in at least two major cities in the region by 2010 Reduce ambient fine particulate concentrations by 10 percent in two major cities by 2005 and by at least 20 percent in four major cities by 2010 Leaded gasoline phased out in major cities of half the countries by the year 2005 and in all countries by 2010 Diesel and gasoline quality improved in at least two countries by 2005 and in all countries by 2010	**Short term (FY 2000–03)** Promote the formulation of urban air quality action plans in at least two major cities of the region. Promote clean fuels and urban air quality improvement in at least two Bank operations Promote the use of natural gas in three large cities of the region by assisting with regulatory reform and infrastructure development Promote the phase-out of leaded gasoline in major cities of at least two countries in the region through policy studies and dialogue **Long term (FY 2004–08)** Mainstream air quality action plans and urban air quality improvement projects Mainstream the use of natural gas for residential/commercial, industry, and power use through (a) the development of transmission and distribution networks and (b) reform of the hydrocarbon sector by establishing competitive markets and eliminating price distortions	Urban environment project in Bangkok (FY 2000), focusing on air quality improvement. One or more air pollution control projects in China for FY 2001–02. Boiler renovation and coal-to-gas boiler conversion in Beijing (FY 2000). West Java gas distribution project as a component of the energy SECAL for FY 2001. IFC and Bank guarantee support of combined cycle gas plant in Vietnam. Proposed gas distribution project in China for FY 2001 Leaded gasoline phase-out initiative in Vietnam to be piloted in Hanoi and Ho Chi Minh City

Strategic Objectives	Outcomes	Actions Needed	FY 2000–02 Outputs
Promote environmentally sustainable production and use of energy resources	Significant progress in improving efficiency in production, transmission, and consumption of energy Increased use of economically viable renewable energy for grid and off-grid applications	**Short and medium terms (FY 2000–03)** Support sector reform in the power industry to encourage efficient expansion and operation of generation capacity Support regional energy trade Implement DSM projects, reduce power transmission and distribution losses, and promote non-utility energy conservation in at least two countries Where feasible, rehabilitate coal-fired thermal power plants to improve energy efficiency and meet environmental standards Support renewable energy investments, including rehabilitation of hydropower facilities through establishment of policies, improving planning capabilities, strengthening private/public sector capacity, and improving access to financing **Long term (FY 2004–08)** Improve significantly energy efficiency in power grid systems as well as by industrial and residential users Continue expansion of the use of renewable energy	Power sector reform initiatives in several countries, including China and Vietnam Sector work on regional energy trade in the Greater Mekong region Prepare a hydrocarbon sector reform note for Indonesia and a power sector strategy paper for Cambodia Power sector rehabilitation loan for China, and a system improvement and energy efficiency project in Vietnam Thailand: DSM component in power distribution projects continues. An industrial ESCO project for energy efficiency improvement Grid and off-grid renewables projects in China, Vietnam, the Philippines, Lao PDR, Cambodia, and possibly Indonesia
Mitigate the potential impact of energy use on global climate change	Declining trend in GHG emissions from projects financed by the Bank, relative to cases without Bank financing, through a combination of: ■ development of cleaner sources of energy (hydro, other renewables, natural gas) ■ improve efficiency of energy generation and end-use consumption	**Short and medium term (FY 2000–03)** Continue to promote energy efficiency and renewable energy development, including new GEF-supported operations in four to five countries. All activities to emphasize market development and strengthening of institutional and regulatory frameworks **Long term (FY 2004–08)** Mainstream energy efficiency/DSM investments in all Bank energy operations and in two to three additional large-scale renewable energy development projects. Have in place policy, institutional, and regulatory framework to implement market-based cleaner energy production and end-use efficiency projects	World Bank Group/GEF strategic partnership will be piloted in China after FY 2000 for wind and small-scale hydro. Phase 2 GEF-supported Energy Conservation project in China further promoting ESCO development DSM/energy efficiency project in Vietnam in FY 2002 Grid-integrated PV pilot project to demonstrate the value of PV in displacing thermal generation and loss reduction (IFC, FY 2000). IFC Renewable Energy and Energy Efficiency Fund and Small and Medium-Sized Enterprises Funds to support renewables and efficiency investments Solar Development Corporation to support photovoltaics investments in East Asia (See other energy efficiency and renewables projects above)

(Continues)

Strategic Objectives	Outcomes	Actions Needed	FY 2000–02 Outputs
Develop capacity for environmental regulation, monitoring, and enforcement across all government levels	Efficient environmental policies and regulations related to energy production and use, transport fuels, and vehicle emissions are enacted and enforced by 2010 in at least half of the countries. Significant progress in enforcement achieved by 2005	**Short and medium terms (FY 2000–03)** In borrowing countries: Training/building local capacity for EAs; strengthening environmental regulations for private sector development (IPPs, etc.); developing monitoring capabilities; improving enforcement capacity of regulatory agency and legal system; support environmental systems of local governments and communities; and promote private-public partnerships and participation for environmental improvement **Long term (FY 2004–08)** Establish a coherent regulatory framework with appropriate standards, regulations, and incentives for improving environmental performance.	Continuing support for national and local environmental regulatory agencies in China and Indonesia and new support for Vietnam. Technical assistance for monitoring, coordination of policies, and public-private initiatives to improve compliance Undertake one to two studies in cooperation with client countries on the relationship between air pollution and health
Make the region more responsive in addressing the adverse environmental impacts of the energy sector	Develop more effective ways of generating and sharing knowledge within the region; develop cooperative and cross-sectoral programs; establish a framework for wider participation among all stakeholders	**Short and medium terms (FY 2000–03)** Improve the sharing and dissemination of information within the region through the relevant thematic groups and through collaborative and interdisciplinary project task teams Establish a cohort staff and expertise in the region for delivering quality and timely services to our client countries in the energy-environment area Enhanced donor coordination to increase the effectiveness of support for renewables and energy efficiency in client countries **Long term (FY 2004–08)** Establish the energy-environment area as a key lending and policy support area of the Bank for countires of the region	Undertake energy-environment reviews in one to two countries Coordination of key sectors within the region (energy, urban, transport, rural/agricultural, environment) responsible for delivering energy-environment projects. Asia Alternative Energy Program to continue to provide project development and implementation support in East Asia Donor coordination for renewables/efficiency investments and capacity building in China and Vietnam will continue and expand to other countries

Annex 7: Monitorable Progress Indicators for the Europe and Central Asia Region

Strategic Objectives	Outcomes	Actions Needed	FY 2000–02 Outputs
Protect human health in urban areas from the adverse effect of air pollution due to fuel combustion in the transport, power, and residential sectors	Reduction of atmospheric particulate concentration by 10–20 percent in three to five large cities Reduction of atmospheric lead concentration by 10–30 percent in at least 10 large cities	In the framework of the regional Clean Air Initiative, demonstrate approaches to cleaning up and preventing air pollution by working with a small number of carefully selected cities on the development and implementation of integrated clean air action plans Facilitate coal-to-gas conversion of district heating boilers, combined heat and power plants, and individual household heating appliances by providing financing and working with governments toward the elimination of price distortions and the commercialization/privatization of gas distribution and marketing Provide technical assistance to facilitate the phasing-out of lead and to increase the market share of unleaded gasoline across the region Facilitate transport fuel quality improvements in Central Asia and the Caucasus through technical assistance	Clean air action plans for three to five cities Implementation of four already approved coal-to-gas conversion projects; preparation and initial implementation of four new projects; privatization and expansion of gas distribution systems with Bank assistance in four countries Refinery assessments in four countries; new limits for lead concentration in gasoline in five countries; and lead awareness programs in eight countries Report with recommended fuel quality improvement actions
Protect natural resources from the adverse impact of water and air pollution due to coal mining, oil production, transport and refining, and electricity generation	Salinity of water bodies around coal mining areas reduced by 20 percent in Poland; water outflow from mining operations reduced by 40 percent in Ukraine, 20 percent in Russia, and 20 percent in Romania	Finance the closure of uneconomic coal mines and site remediation in Poland, Romania, Russia, and Ukraine	Implementation of four already approved mine restructuring projects/adjustment operations and preparation and partial implementation of four new mine restructuring projects
	Avoidance of another major oil spill in the Komi region in Russia and rapid clean-up of oil spills in Romania Environmentally safe development of offshore oil production at four sites	Finance pipeline rehabilitation and the removal of water from crude oil before pipeline transportation in the Komi region of Russia Finance oil spill emergency response equipment in Romania Facilitate the prevention and clean-up of oil spills through technical assistance provided to the government and oil companies in Russia Finance EIAs for offshore oil development projects in Russia and the Caspian	New oil/water separation plant and refurbished pipeline in Komi New oil spill emergency response equipment placed in strategic locations in Romania Report with recommended actions and a public awareness effort EIA reports

Strategic Objectives	Outcomes	Actions Needed	FY 2000–02 Outputs
	Compliance with protocols on long-range transboundary sulfur dioxide and nitrogen oxides pollution	Finance coal-to-gas conversion (see above) and rehabilitation of selected power plants with measures to reduce SO_2 and NO_x emissions	Implementation of 10 already approved coal-to-gas conversion/rehabilitation projects and preparation and partial implementation of five new projects
Mitigate the potential impact of energy use on global climate	Meeting commitments under the Kyoto Protocol to the UNFCCC	Facilitate the reduction of wasteful energy consumption through the financing of investments in energy efficiency; foster the establishment of ESCOs; and encourage governments to correct energy price distortions and to remove subsidies	Implementation of two already approved energy efficiency projects; preparation and partial implementation of two new energy efficiency projects; and establishment/strengthening of autonomous, professional regulatory entities with Bank assistance in at least 10 countries
		Finance investments through GEF to utilize renewable energy sources such as hydropower, biofuels, and geothermal energy	Implementation of three already approved renewable energy projects and preparation and partial implementation of six new projects
		Finance investments, partly through GEF, and provide assistance to reduce the leakage of gas from transmission and distribution lines and to reduce gas flaring	Implementation of two already approved gas transmission/distribution rehabilitation projects; preparation and partial implementation of two new projects; and two reports on measures to reduce gas leakage and gas flaring in Russia
		Finance investments through GEF to capture and utilize coalbed methane in at least two countries	Preparation and implementation of two coalbed methane utilization projects

Annex 8: Monitorable Progress Indicators for the Latin America and Caribbean Region

Strategic Objectives	Outcomes	Actions Needed	FY 2000–02 Outputs
Mitigate the potential impact of energy use on global climate change	Declining trend in GHG emissions from projects financed by the Bank, relative to cases without Bank financing. To be achieved by implementing a combination of: ■ development of cleaner sources of energy (including hydropower and other renewables and natural gas) and, where economically feasible, substitution of dirty fuels ■ improvement of energy generation efficiency and end-use consumption, including industry, transport, households	**Short and medium terms (FY 2002–05)** Agree with governments in at least four countries to invest in energy efficiency/DSM projects and in at least one country to participate in GEF-financed demonstration of large-scale renewables. Activities will emphasize market development and strengthening of institutional and regulatory frameworks Improvement in interconnection of electrical grids to facilitate trade **Long term (FY 2008)** Agree with additional two to three countries on energy efficiency/DSM investments and one additional large-scale renewables demonstration. Have in place policy, institutional, and regulatory framework to develop and implement cleaner energy production and end-use projects	Bank operations in at least four countries on energy efficiency/DSM/on-grid renewables Consultation/dissemination workshops in at least three countries in the region to discuss models for alternative energy commercialization Sector reports and EERs characterizing the markets and opportunities for Bank assistance on efficiency/on-grid renewables/environmental mitigation in at least five countries Integration of strategies to deal both with local airborne pollutants and with GHG emissions in at least two countries
Improve access to modern energy by rural populations to alleviate poverty, reduce health impacts of traditional fuel use	Significant progress in household access to cleaner commercial energy Increase share of cleaner commercial fuels by 10 percent in rural sector of at least one country by 2005 and by 20 percent in at least two countries by 2010	**Short and medium terms (FY 2002–05)** Examine rural energy options in countries of the region and agree with at least three countries on rural electrification program, with emphasis on off-grid and dispersed populations. Quantify to extent possible health and other benefits of substitution of new technologies for traditional fuels. Create strategies to promote the use of off-grid power-generating technologies **Long term (FY 2008)** Mainstream energy access to rural communities and urban poor in Bank operations in at least three more countries; and in other countries develop consensus with governments on improved rural energy access	Bank operations in at least three countries on rural electrification, with renewable energy components for off-grid/isolated areas Consultation/dissemination workshops in at least three countries in the region to discuss models for delivery of basic energy services to rural areas Sector reports characterizing the rural markets and identifying opportunities for Bank assistance on rural electrification in at least four countries

(Continues)

Strategic Objectives	Outcomes	Actions Needed	FY 2000–02 Outputs
Improve air quality in urban areas	Air quality (particulate matter and ozone) improved in at least two major cities in the region by 2010; most other major cities to adopt similar programs Leaded gasoline phased out in all countries by 2005 Diesel and gasoline quality improved in at least two countries by 2005 and in all countries by 2010	**Short and medium terms (FY 2002–05)** Consolidate dialogue with the local and national governments of all major cities in the region within the framework of the Clean Air Initiative sponsored by the Bank, such that major air quality programs are established or improved. Encourage government commitment to implement a strategy to control urban air pollution in at least four major cities, demonstrated by prior actions, e.g., by establishing a coordinating body for implementing a multisectoral pollution control program. Facilitate the restructuring of the petroleum sector, improving fuel specifications and pricing policies. Introduce cleaner vehicles and transport demand management measures and introduce measures oriented to reduce and mitigate emissions of particulate matter **Long term (FY 2008)** Have comprehensive air pollution management programs in place at least in two major cities and start introducing similar programs in at least two other major cities	Establish Bank multisectoral teams Workshops on cleaner vehicle engines/emission standards, fuel specifications, pollutant emission, and concentration monitoring and information tools, modeling capacity, verification programs for different vehicle fleets, and cleaner and more efficient use of energy in industries and services, in at least two countries Bank operations to support the development of a comprehensive clean air program in at least two countries
Improve capacity for environmental regulation and enforcement	Enactment and enforcement of sound environmental policies and regulations related to energy production and use, including transport fuels and equipment, urban planning, and ecological control of biogenic sources of airborne pollutants Improved air quality management; greater and more integrated role of local authorities and civil society	Develop the necessary institutional capacity to manage comprehensive air quality programs; establish an adequate regulatory framework and enforcement measures; improve monitoring and information systems, modeling, and planning capacity; and facilitate the improvement and enlargement of local scientific base Promote greater participation and awareness of local-level government, communities, and private stakeholders	Review of standards and regulatory framework in at least three large cities Improvement in modeling capacity geared to strategic planning and contingent program management in at least four countries Economic analysis of most common measures to improve air quality and development of economic incentives in at least three countries

Annex 9: Monitorable Progress Indicators for the Middle East and North Africa Region

Strategic Objectives	Outcomes	Actions Needed	FY 2000–02 Outputs
Improve health and well-being of urban residents through improving ambient air quality in highly polluted cities	Air quality improved in at least two major cities in the region by 2010; most other major cities adopt similar programs Leaded gasoline phased out by 2005 in three countries Diesel quality improved in at least two countries by 2005 and in all countries by 2010 Fuel switching to cleaner technologies involving gas, including co-generation	**Short term (FY 2002)** Consensus within the Bank and with respective governments on the basis for reducing air pollution in selected cities; continued commitment to phase out leaded gasoline in at least two countries; advance dialogue leading to commitment to phase out lead in all countries; pursue implementation of economic-based pricing for energy **Medium term (FY 2005)** Pursue government commitment in all countries to implement a strategy to control urban air pollution demonstrated by prior actions, e.g., by switching to cleaner fuels in thermal power plants, improving fuel specifications and pricing policies, and continued commitment to phase out leaded gasoline **Long term (FY 2008)** Have comprehensive air pollution control programs in place at least in two major cities and start introducing similar programs in two other major cities	Regional energy and environmental strategy of 1995 updated Undertake Energy and Environment Review (EER) in two countires Integrate energy and environment considerations into the CAS in at least one country At least three new power generation projects based on gas Continuing DSM for improving energy efficiency in at least one country Undertake energy pricing studies in at least three countries
Alleviate poverty and protect health of rural and poor urban families through reducing exposure to indoor air pollution from the use of traditional fuels	Significant progress in household access to cleaner commercial energy Increased share of commercial fuels by 10 percent in at least one country by 2005 and by 20 percent by 2010 Increased role of renewable energy technologies	**Short term (FY 2002)** Start the examination of rural energy options and build consensus in at least one country on the need for reducing indoor air pollution; exchange ideas with other donors on the most practical ways of addressing the indoor air quality problem **Medium term (FY 2005)** Mainstream energy access to rural communities and urban poor in Bank operations in at least two countries; in other countries, secure firm government commitment to improved rural energy access **Long term (FY 2008)** Facilitate government commitment to take concrete steps toward addressing the indoor air pollution problem in at least one country, as evidenced by the agreement in principle of at least one government to integrate indoor air pollution in CAS and to include rural energy access components in power sector restructuring projects	Sector reports that analyze the health impacts of indoor pollution for at least one country Identify main options to improve rural energy access in at least two countries where rural access is low compared to others in the region (technical, institutional, and financing)

(Continues)

* Prepared by MNSID and MNSRE

Strategic Objectives	Outcomes	Actions Needed	FY 2000–02 Outputs
Develop capacity for environmental regulation, monitoring, and enforcement across all levels of governance and civil society	Sound environmental policies and regulations related to energy production and use and to transport fuels are enacted, enforceable, and enforced Improved environmental governance; greater role of local authorities and civil society Stringent emissions standards for new transport vehicles	**Short term (FY 2002)** Training and building local capacity for EAs; strengthening environmental regulatory framework, monitoring, and enforcement functions; improvement of enforcement capacity of regulatory agency and legal system **Medium term (FY 2005)** Improved standards, regulations, and incentives for internalizing externalities; improved quality of economic analysis of environmental impacts for planning, investment decisions, and policy purposes; facilitate improved understanding of cost benefits of environmental improvement **Long term (FY 2008)** More support to local level governments and communities, promote private-public partnerships and participation	Assist in the establishment of adequate environmental management capacity in at least one country Include technical assistance component in projects Use IDF or other grant financing for strengthening capacity of institutions on a regional basis
Promotion of environmentally sensitive natural resource exploitation	Environmentally sound exploration and production of oil and gas Rehabilitation and clean-up of existing and decommissioned oil production facilities Reduction in gas flaring	Initiate dialogue with major oil producer countries; establish best practices in pollution abatement and response	Undertake assessments in at least one country through EER work

Strategic Objectives	Outcomes	Actions Needed	FY 2000–02 Outputs
Mitigate the potential impact of energy use on global climate change	5–10 percent reduction in cumulative GHG emissions in three countries by 2010 by comparison with business-as-usual scenario, through the implementation of a combination of: ■ economic pricing and energy subsidy removal to encourage energy efficiency ■ development of cleaner sources of energy (e.g., gas) and, where economically feasible, substituting for high-carbon fuels Increase the volume of energy trade between at least three countries by 2005 Doubling of power generation through renewable energy sources in at least two countries by 2005	**Short term (FY 2002)** Continue to build consensus on the need for power sector reform and promote the removal of tariff distortions; reach agreement with at least two governments to implement pilot efficiency/DSM component in power sector restructuring projects; strengthen relationships with strategic partners to develop alternative energy programs; propose targeted interventions in at least one country to promote the development of renewable energy technologies; carry out analytical work and share information with other regions on energy trade **Medium term (FY 2005)** Agree with government in at least one country to promote at least one renewable energy development project and to pursue institutional development of borrowers to launch renewable energy projects and energy efficiency programs; reach agreement in principle with at least two governments on the need for promoting regional energy trade; examine Bank support (investment lending, guarantees, advisory services) for energy trade projects such as gas **Long term (FY 2008)** Have in place policy, institutional, and regulatory framework to develop cleaner energy projects; mainstream advisory services to promote energy trade and renewable energy projects	Information sharing and consensus building workshops on power sector reform, energy trade, and alternative energy options Energy pricing studies in at least three countries Sector reports demonstrating the economic benefits of energy trade Agreed strategy with one to two countries on promoting alternative energy projects Agreed gas development strategies with at least two countries Promote regional power and gas interconnections

Annex 10: Monitorable Progress Indicators for the South Asia Region

Strategic Objectives	Outcomes	Actions Needed	FY 2000–02 Outputs
Alleviate poverty and protect health of rural and poor urban families by reducing exposure to indoor air pollution from the use of traditional fuels	Significant progress in household access to cleaner commercial energy Increase share of cleaner commercial fuels by 5–10 percent in at least one country/state by 2005 and by 20 percent in at least two countries/states by 2010	**Short term (FY 2002)** Start the examination of rural energy options and build consensus in at least one country on the need for reducing indoor air pollution; exchange ideas with other donors on the most practical ways of addressing the indoor air quality problem; reach agreement with at least one government to integrate indoor air pollution in CAS preparation **Medium term (FY 2005)** Obtain commitment of at least one government to take concrete steps toward addressing the indoor air pollution problem and to include rural energy access components in Bank projects **Long term (FY 2008)** Mainstream energy access to rural communities and urban poor in Bank operations in at least two countries/states; and in other countries/states obtain firm government commitment to improved rural energy access	Review of Bank experience with indoor air pollution Awareness workshops Sector reports for at least one country/state that analyze the health impacts of indoor pollution; identify options (grid and decentralized) to improve access to commercial fuels in poor communities; identify interventions and implementation strategy, including local institutional strengthening to address indoor air pollution problems; and define a framework for providing commercial fuels to rural communities and poor urban families Wider participation of NGOs and civil society
Improve health and well-being of urban residents by improving ambient air quality in highly polluted cities	Significant improvements in air quality in at least two major cities in the region by 2010; most other major cities adopt air quality improvement programs Reduced use of leaded gasoline in most countries by 2005 Diesel quality improved in at least two countries by 2005 and in all countries by 2010	**Short term (FY 2002)** Reach consensus with respective governments on the basis for reducing air pollution in selected severely polluted cities (such as Dhaka, Delhi, Kathmandu, Lahore, or Lucknow); commitment to phase out leaded gasoline in at least two countries; and advance dialogue leading to commitment to phase out lead in all countries **Medium term (FY 2005)** Facilitate government commitment to implement a strategy to control urban air pollution in at least two major cities, demonstrated by prior actions, for example, by establishing a local coordinating body for implementing multisectoral pollution control program Facilitate the restructuring of the petroleum sector, improving fuel specifications and pricing policies Obtain agreement of all countries to phase out leaded gasoline **Long term (FY 2008)** Have comprehensive air pollution control programs in place in at least two major cities; introduce similar programs in at least two other major cities	Establish Bank multisectoral teams Country and/or regional activities on lead phase-out, cleaner vehicle engines/emission standards Memorandum of understanding with government on air pollution abatement program Agreed issues paper on urban air pollution Clean transport fuel options study in at least two countries and rapid assessment of at least two highly polluted cities, to include identification of the key sources of urban air pollution and an analysis of options for reducing pollution Review and disseminate information about the two-stroke initiative (under preparation) and lessons learned from the Dhaka Air Pollution Learning and Innovation Loan

Strategic Objectives	Outcomes	Actions Needed	FY 2000–02 Outputs
Protect health, prevent degradation of natural resources, and foster productivity by controlling adverse environmental impacts of coal-power development in India	Air and water pollution and land degradation from coal-power development in India reduced, evidenced by (a) progress toward implementing power sector reform in three to four states by FY 2008; (b) doubling of ash utilization from the current level of 2–3 percent in two Indian states by FY 2008; (c) decline in the share of hydropower in power generation mix and increase in use of gas; and (d) rehabilitation of selected degraded areas (e.g., Singrauli) by FY 2008	**Short term (FY 2002)** Disseminate the findings of the India Environmental Issues in the Power Sector (EIPS) study to at least four Indian states and reach agreement with at least two state governments to undertake some form of power sector planning that considers environmental impacts; progress in implementing coal washing; reach consensus on ash management and utilization strategies in at least one large plant; and examine fuel substitution options **Medium term (FY 2005)** Obtain the commitment of at least two state governments in India to increase the utilization of ash; make progress toward initiating the preparation of a few environmentally and socially benign hydropower projects **Long term (FY 2008)** Implement ash management, disposal, and utilization programs; prepare projects to achieve a more balanced energy mix, including greater roles for hydropower and gas; and prepare operations to rehabilitate selected areas degraded through coal-based power development (e.g. Singrauli)	Case study of the Singrauli area EIPS dissemination workshops in Andhra Pradesh, Uttar Pradesh, Rajasthan, and Karnataka Agreed strategy with authorities for addressing environmental and social problems of coal mining operations Assess long-term options for fuel substitution in the context of work on regional energy trade
Develop capacity for environmental regulation, monitoring, and enforcement across all levels of governance and civil society	Enactment and enforcement of sound environmental policies and regulations related to energy production and use and transport fuels Improved environmental governance, including greater participation of local authorities and civil society	**Medium and long terms (FY 2006)** Continue the institutional development of Coal India, including strengthening its capacity to address environmental and social problems Training and building local capacity for EAs; strengthening environmental regulatory framework for private sector development (IPPs, etc.); strengthening monitoring and enforcement functions; improved standards, regulations, and incentives for internalizing externalities; improved quality of analysis of environmental impacts for planning, investment decisions, and policy purposes More support to local-level governments (systematic management training) and communities; promote private-public partnerships and participation	Develop an actions-oriented management training program for regulatory authorities Intensive advisory activities by Bank staff Capacity-building components in Bank projects Outreach and communication strategy

Continues

Strategic Objectives	Outcomes	Actions Needed	FY 2000–02 Outputs
Mitigate the potential impact of energy use on global climate change; position clients to better utilize resources such as GEF	5–10 percent reduction of cumulative GHG emissions, measured in comparison with business-as-usual scenario, in at least two countries/states. To be achieved through the implementation of a combination of: ■ power sector reform in three to four Indian states and power sector reform in at least two other countries by FY 2008 ■ development of cleaner sources of energy (e.g., hydropower; gas) and, where economically feasible, substitution of dirty fuels ■ increases in the volume of energy trade between at least two countries by FY 2008 ■ doubling of power generation through renewable energy sources in at least two countries by FY 2008	**Short term (FY 2002)** Continue to build consensus on the need for power sector reform and promote the removal of energy price distortions; reach agreement with at least two governments (central or state) to implement pilot efficiency/DSM component in power sector restructuring projects; strengthen relationships with strategic partners to develop alternative energy programs; propose targeted interventions in at least one country/state to promote the development of renewable energy technologies; carry out analytical work and learn from the experiences of other regions in energy trading **Medium term (FY 2005)** Agree with governments in at least one country/state to promote at least two renewable energy development projects and to pursue institutional development of borrowers to launch renewable energy projects and energy efficiency programs; reach agreement in principle with at least two governments on the need for promoting regional energy trade; examine Bank support (lending, guarantees, advisory services) for energy trade projects (e.g., hydropower; gas) **Long term (FY 2008)** Have in place policy, institutional, and regulatory framework to develop cleaner energy projects, such as hydropower and gas, for domestic use and for exports in at least two countries; advance dialogue with other governments; and mainstream advisory services to promote energy trade and renewable energy projects	Continue the implementation of power sector reform projects and make progress in developing at least one more (with efficiency/DSM component) Information sharing and consensus building workshops on power sector reform, energy trade, and alternative energy options Sector reports demonstrating the economic benefits of energy trade Agreed strategy with at least two country/state governments on promoting alternative energy programs Agreed hydropower development strategy with at least one government